David Singmaster
4 Jan 2011

Science and Society in Ireland

SCIENCE AND SOCIETY IN IRELAND

The Social Context of Science and Technology
in Ireland, 1800–1950

Edited by

Peter J. Bowler & Nicholas Whyte

The Institute of Irish Studies
The Queen's University of Belfast

Published 1997
The Institute of Irish Studies
The Queen's University of Belfast

This book has received financial assistance under the Cultural Traditions Programme of the Community Relations Council which aims to encourage acceptance and understanding of cultural diversity.

© Copyright reserved.

The contributors assert the moral right to be identified as the authors of their papers.

All rights reserved. No part of this publication may be reproduced, stored in a retrieval system, or transmitted, in any form or by any means, electronic, photocopying, recording, or otherwise, without prior permission of the publisher.

ISBN 0 85389 669 0

Printed by W. G. Baird Ltd, Antrim

Front cover: Lord Rosse's great telescope at Birr Castle, from an engraving in Sir Robert Stawell Ball's *The story of the heavens*, London, 1893.

Contents

	Page
Editors' Introduction	vii
Roy MacLeod **On Science and Colonialism**	1
David Attis **The Social Context of W. R. Hamilton's Prediction of Conical Refraction**	19
James Bennett **Science and Social Policy in Ireland in the mid-Nineteenth Century**	37
Nicholas Whyte **Science and Nationality in Edwardian Ireland**	49
Hugh Torrens **James Ryan and the Problem of Irish 'New Technology' in British Mines in the early Nineteenth Century**	67
W. Garrett Scaife **Technical Education and the Application of Technology in Ireland, 1800–1950**	85
Richard A. Jarrell **Some Aspects of the Evolution of Agricultural and Technical Education in Nineteenth-Century Ireland**	101
John Wilson Foster **Natural History in Modern Irish Culture**	119
Anne Buttimer **Twilight and Dawn for Geography in Ireland**	135
Sean Lysaght **Science and the Cultural Revival, 1863–1916**	153
Index of People and Institutions	167

Editor and Contributors

Peter J. Bowler is Professor of History and Philosophy of Science at The Queen's University of Belfast.

Roy MacLeod is Professor in the Department of History, University of Sydney, Sydney NSW 2006, Australia.

David Attis is a graduate student in the history of science in the Graduate College, Princeton University, Princeton NJ 08544, U.S.A.

Dr James Bennett is Keeper of the Museum of the History of Science, University of Oxford, Broad Street, Oxford OX1 3AZ.

Nicholas Whyte is completing a doctorate in History and Philosophy of Science at The Queen's University of Belfast

Dr Hugh Torrens lectures in the Department of Geology, Keele University, Keele, Staffordshire ST5 5BG.

W. Garrett Scaife is Professor in the Department of Mechanical and Manufacturing Engineering, Trinity College, Dublin.

Richard A. Jarrell is Professor in the Department of Science Studies, Atkinson College, York University, 4700 Keele Street, North York, Ont. M3J 1P3, Canada.

John Wilson Foster is Professor in the Department of English, University of British Columbia, 397–1873 East Mall, Vancouver, B.C. V6T 1Z1, Canada.

Anne Buttimer is Professor in the Department of Geography, University College Dublin.

Dr Sean Lysaght lectures at the Regional Technical College, Galway, Castlebar Campus, Co. Mayo.

Editors' Introduction

The papers in this volume were originally presented at a conference on 'The Social Context of Science, Technology and Medicine in Ireland, 1800–1950' held at the Royal School Armagh, on 28–29 October 1994. The papers on the history of medicine are to be published separately in a volume edited by Professor Greta Jones and Dr Elizabeth Malcolm. Thanks are due to the Royal Society, the British Academy, the Wellcome Trust and the Cultural Traditions Group for the funding that made it possible to bring an international group of speakers to Armagh.

The purpose of the conference was to create a forum in which the growing interest in the history of science and technology in Ireland could be expressed, especially in a way that could transcend the narrow (or what was once called the 'internalist') approach to the history of particular technical disciplines. Academic historians of science and technology have long recognized that – for all the assumed internationalism of science – practice has in fact been constrained and influenced by local political, economic and cultural circumstances. Science and engineering are parts of the social life of the population in which they are conducted. Associated with this movement within the history of science has been a growing interest in the national identity of scientific communities. Science and technology have developed differently in Britain, America, France or Canada (to give some well-studied examples) because each of these countries had to develop its own way of organizing technical expertise within its governmental, industrial and cultural institutions. Comparative studies based on national differences have increasingly been used to throw light on the factors which influence the development of science and technology in their social environment.

It was within this context that the Armagh conference was planned. Ireland is a small country, but it has a unique social history and a legacy of significant contributions to scientific and technological development. Many scientists and engineers working in Ireland have acquired interest and expertise in the local history of their own fields. Significant products of this interest include Gordon Herries-Davies' *Sheets of Many Colours* (now superseded by his *North from the Hook: 150 Years of the Geological Survey of Ireland*), and Patrick Wayman's *Dunsink Observatory, 1785–1985*. Unfortunately, this dimension of the Irish experience has been largely neglected by social historians, at least until recently, and there are few trained historians of science on the island to offer a lead. Several historians from other countries – some of them represented in this volume – have taken an interest in Irish science precisely because it provided them with a tool for the kind of comparative studies that have proved so useful elsewhere. It was hoped that by presenting the work of historians from all these backgrounds together at one conference, a fruitful dialogue would ensue and enrich the future study of the history of science and technology in Ireland. Scientists and engineers with local knowledge may learn from the sophisticated techniques developed by historians of science studying parallel problems in other countries. Irish social historians may also be encouraged to see that science and technology are not foreign topics that can safely be excluded from their studies.

In the papers that follow, Professor Roy MacLeod, the conference's keynote speaker, provides an overview of the general theme of science and colonialism, a topic which raises obvious questions in light of the Irish experience. Professor MacLeod has written extensively on the theme of science and European expansion, and with Richard Jarrell has recently co-edited *Dominions Apart* (issued as parts 1 and 2 of *Scientia Canadensis*, vol. 17), a collection of papers comparing the roles played by science in Canadian and Australian history. The other speakers address more specific topics in the history of Irish science and technology, all relating their subject matter to local social and cultural concerns. David Attis shows that Sir William Rowan Hamilton's work in mathematical physics – although at first sight irrelevant to any specifically Irish experience – was nevertheless seen by Hamilton himself as something that would affect perceptions of Irish cultural life. James Bennett's name is already familiar through his account of the Armagh Observatory, *Church, State and Astronomy in Ireland*. Here he addresses the wider theme of science and social policy in Ireland in the mid-nineteenth century. Nicholas Whyte, who is now completing a doctoral dissertation on the social context of science in Ireland in the late nineteenth and early twentieth centuries, shows how nationalism affected the development of science in the Edwardian period.

At a more down-to-earth level, Hugh Torrens notes how social perceptions limited the spread of new mining techniques developed in Ireland. Dr Torrens has written extensively on the practical aspects of the history of geology, and here applies his experience to a topic with a significant Irish dimension. Garrett Scaife, professor of mechanical and manufacturing engineering at Trinity College, Dublin, surveys technical education in Ireland from 1800 to 1950. He argues that – because central government played a more active role in nineteeth-century Ireland – technical education here followed a pattern more common in France than in England. Richard Jarrell has written several articles on science in Ireland, including one which compares the Irish attitude to science with that manifested in Quebec (where there was also a large Catholic population under British government). Here he looks at agricultural education in nineteenth-cen-

Editors' Introduction

tury Ireland, noting how over-theoretical educational programmes had little popular appeal.

Despite his academic base in Vancouver, John Wilson Foster's name is familiar to many readers in Ireland. He has recently become interested in the role played by natural history in Irish culture, and has co-edited a book soon to be published on the topic (with Helena Chesney). His paper provides a sometimes controversial look not only at changing attitudes to natural history in Ireland, but also at our modern perception of what counts as Irish natural history. Anne Buttimer has written extensively on the history of geography (her book, *Geography and the Human Spirit* appeared in 1993). Her article uses the concept of cultural 'gatekeeping' – defining a body of culture as the property of a particular social group – to compare the role of geography in Ireland and America. Finally, Sean Lysaght looks at science and the cultural revival, noting how natural history especially played a role in the cultural life of late nineteenth and early twentieth-century Ireland.

The articles collected here thus range over a wide area of the history of Irish science and technology. All are based on the assumption that history is something more than the gathering of information about the past: the information has to be interpreted to give it meaning. Some of them challenge popular perceptions, including the all-too-frequent dismissal of science as irrelevant to the native Irish experience. Some raise what could still be controversial issues, such as the tendency to sideline developments that have taken place north of the border. The history of science and technology, like other aspects of Irish history, may not be without its controversies. The contributors to this volume do not always agree with one another, as in the case of Foster and Lysaght on the role of the cultural revival. It is precisely to show that this area offers exciting fields of study in which intepretations are still being thrashed out that these papers are published. If they stimulate further study and debate, they will have served their purpose.

Peter J. Bowler
Nicholas Whyte

Chapter 1

On Science and Colonialism

Roy MacLeod

INTRODUCTION[1]

This meeting comes – I need hardly remind anyone here – at an extremely auspicious moment in the long history of colonialism on this island, a time when discussion is replacing violence, when many of the polarisations and ambivalences that have marked Ireland's passage into the modern world are becoming instead the subjects of reflection and hope. Ireland's past, no less than Britain's past, shapes the framework of present debate – and if it is true, which maybe some will deny, that the English never remember history, while the Irish forget none – then we have every reason to recall the place of science and culture both in Ireland and between two countries so forcibly tied by geography, migration, economics, and an Act of Union. That place remains to be flexibly defined, in a context that is no stranger to fixed opinions. Since 1800, Protestant historians have sedulously qualified romantic conceptions of a pre Norman Ireland, while their nationalist and Catholic counterparts have striven to defend them.[2] And just as Irish history has seen so much scholarly dissension over the anatomised body of the colonial past, so the relations between colonialism and science have been construed as a metonym of dependence, its resolution involving a declaration of independence, a demand for sovereignty, in a post-colonial world.[3] Looking at the imperial map, on the one side, it is easy to picture the subject agencies – black, Muslim, or Irish; and the apologists, WASP, Oxbridge, reflecting traditions of deference and the Crown, assisted by a bourgeois commercial and entrepreneurial middle class ethos that sought unity in progress from palm to pine.[4] But these stereotypes are not impermeable, nor are they permanent, and the traffic in ideas and influences that we now perceive is not linear, but two-way.

They are also lasting, and if analogous images endure in the colonial narratives of Iberia, France, Germany, and the Netherlands as well, they are a sure sign that Britain was not alone in reflecting the unwisdom of forgetting colonisers, for so long as the colonised remember.

And, of course, the colonised never forget. This fact raises an issue of importance for scholars in the history of science learning from scholars of empire. Too often, the struggle against colonialism has been cast in terms of a timeless, transcendent, ahistorical process, devoid of a developmental or sequential view of events in which rights are established by passage of time. Every generation is equidistant from eternity, said Ranke, as a warning against those who, in Oliver MacDonagh's splendid phrase, 'follow the furrow of progress to the present, and praise the dead ploughman who deviates least from the appointed line'.[5] And European historians of science, working within the western tradition, on the whole have favoured a view of natural knowledge that is sequential, and in some sense, progressive – whether evolutionary or revolutionary, realist or socially constructed we may at leisure debate. This is why Whiggism is for us such a constant, if uncomfortable, companion. It is also one reason why historians of science fail to meet our colleagues in colonial history on common ground. For while our knowledge of nature must in some sense be progressive – of a nature that is itself transcendent, and timeless – we are required to situate science and ourselves in a context that is culturally and historically contingent. Insofar as we are thus 'transcendents' caught in time, we share the general historian's preoccupation with space and place. And it is to these categories, and our sense of them, that my remarks are directed.

While it would be incorrect to exaggerate our self-importance, the growing interest in 'science as nationality' represented by this conference does seem to reflect a growing sense of maturity and post-colonial self-confidence in Ireland similar to that found in 1976 by the Americans, and later in Canada, Australia, and New Zealand.[6] However, this becomes a matter not merely of joining a queue, eager to show a rich share of Irish contributions to currents of intellectual development. We cannot expect the world to remember, if we never forget, that both Sir William Hamilton and the Duke of Wellington were Irishmen. Instead, we have an obligation to our post-colonial sensibility, to show how, and why, a particular sense of place, space and time, contributed to what is distinctive about science in Ireland, which we may by extension apostrophise as 'Irish science'. Space, time, place, nationhood – these are the elements in the newer historiography that are uniting the history of science and colonial history.

This new historiography is important for another reason. If we today see ourselves at a crossroads in Ireland, and in Irish history, elsewhere one sees histories passing at the busy intersection between political history, the history of science, and the history of Europe's encounter with 'overseas'. The 'expansion of Europe' leads us to a broad interchange dominated by globalisation, the rise of East Asia, the decline of deference and traditional power structures, and vast movements of endangered peoples and species. Historians traverse a thoroughfare criss-crossed by exits pointing to environmental conservation, regional self-determination, and community self-knowledge – warned by signs that force us to consider the world not only as historians, but as citizens as well. The complex process of 'transition' taking place in and between postcolonial cultures – whether east and west, or north and south – has wrought huge changes in our perception of the colonial relationship, in culture and politics as in class, race and gender.[7] The impossible task of distinguishing between 'ourselves' and the 'other' – a conceit recognised by anthropologists long ago, if still very difficult for many historians to grasp – has reminded us how uncomfortably close the colonial world and the Cold War coexisted, and how neither are far removed from the mentalities of the present.[8]

The challenge, I suggest, is to associate more closely what we do as historians of science with what is being done by re-reading the narratives of colonial history. Whether as colonisers or colonised, historians are repositioning themselves in relation to colonial discourse theory. It does not surprise us, in the shadow of Heisenberg, that observations are affected by observers; yet many persist in thinking that the natural environment within which natural science develops has somehow left science untouched. Fortunately if the affiliation of ideas is demonstrable in the history of ideas, so it is in the history of place, where individuals and environments condition the ways in which questions are put, answers are derived, institutions established, and traditions maintained.[9]

In the following pages, I want to emphasis the importance of certain key ideas in the history of colonial science, and in its changing historiography, and to point to the potential of these changes for a new historiographical agenda, one which is re-examining categories, juxtaposing cultures and contemplating the possibility of not just one history of science and colonialism but of several. As you will see, this agenda reflects my interest in three particular dimensions – the history of *mentalities*, of scientific styles and strategies, and of the sciences of tropical nature, and of the 'south' metaphorical and physical.[10]

We begin on common ground. Science has no nation; but nations have science, and in the era of European imperialism beginning in the last quarter of the nineteenth century; there arose a interest in making science serve the interests of imperial efficiency and colonial development. The interest was fresh – but hardly new – as for over two centuries, the nations of western Europe, initially led by Iberia, sought out posts for commerce and Christian settlement.[11] Whatever their motives and means and whatever their allegiances to economic doctrine, western Europeans came to be united on the principle by which colonies served as plantations or primary producers for the trade and manufacturing industries of the metropolis. Within this context grew, by the early nineteenth century, a diverse range of rational projects which we collectively label 'colonial science', and which have during the last decade been situated in the colonial projections of Spain, Britain, France, and Germany.[12] These persisted, albeit with important modifications, until the end of the second world war.

Today, the history of this engagement is sufficient to attract an international audience. Once upon a time, this was not so. Indeed, until the 1970s, selling the history of colonial science to historians of science was like selling a little-known commodity low in a falling market. In the last twenty years, however, for reasons which we may want to examine, the meanings of this process have become increasingly relevant to historians of western science in the Indian sub-continent, the Americas, and what Europeans call the Far East.[13] At the same time, there have emerged several interpretative perspectives, competing for attention. For some, colonial science is a museum of the moving image, in which Europeans are displayed constantly moving south and east. To others, the important narratives are those in which ambitious locals use scientific knowledge to help frame a response to the external domination of a world culture. What is problematic about this process has been the subject of international conferences in 1981 in Melbourne and 1990 in Paris – where historians exposed a huge range of colonial experiences, and reflected a variety of cultural styles in narrative scholarship.[14]

Following the pioneering work of Donald Fleming, George Basalla, and I.B. Cohen,[15] we have seen diffusionist models rise and fall. The concept of 'diffusion' itself, first conveying a sense of unassertive benevolence, became less friendly when recast in the more aggressive language of insemination and irradiation. In its place, came a plethora of case studies, relocating the imperial expansion of western scientific, medical and technological doctrines in shaping the form and content of colonial institutions.[16] In the history of colonial medicine, we have traced the medicalisation of the knowable world, the incorporation of European theory and practice into alien settings, and the process by which Hippocratic ideologies of medical environmentalism also 'tropicalised' Europe.[17] In the history of colonial science, we have seen how European debates confront, assimilate, and in turn become transformed by the floral, faunal, mineral and material conditions of existence in the colonial world.[18] In the history of technology, we have seen how the tools of conquest that characterised early colonial settlement became transformed by the cultures they conquered.[19] In the process of discovery, we have learned that the 'centre-periphery' simile is an artefact to be explained, and not a truth to be assumed.[20] We know that the language, methods and projects of empire became means both of enlarging western knowledge, and as civilising practice.[21] The legacies of the Crystal Palace, incorporated in the metropolitan museums and international exhibitions of the nineteenth century, left their impression on the world at large, representing indigenous people as 'material cultures' in themselves.[22] Civilising missions and practical men, within different imperial systems, appeared in the settler colonies as representatives of an ideology *sans doctrine*, flexibly adopting and reinventing languages in which science is made to suit the interests of both the coloniser and the colonised. Slowly, and by degrees, colonial science metamorphosed and became appropriated as an instrument of colonial nationalism and self-identity.

Within the last five years, the interrogation of this past has become more confident and sophisticated. We see the discourse of interest – so well established in the sociology of scientific knowledge – implicit in the mechanisms by which science and technology have been transferred or withheld, assimilated or rejected, indigenised or monopolised. We have looked beyond globalist world systems and modernisation theories bred of the Chicago School and Immanuel Wallerstein, as well as neo-classical and Marxist theories that explained away imperialism as a residual expression of late modern capitalism. We have dismissed explanatory models that have minimised the significance of local cultures and trivialised local initiatives in the process of assimilation. 'Centres of calculation', in Bruno Latour's memorable phrase, remain and are privileged, but instead of a few, based mainly in Europe, we now see a multiplicity, whose visibility and influence are determined by much more cosmopolitan patterns of information and recognition. Qualifying the classic formulae of Mertonian sociology, the history of colonial science celebrates diversity as much as universality, and seeks to complement what Lafuente calls the 'omnipowerful and omnipresent metropolis' with a variegated, irreverent, quixotic – rather than exotic – periphery.[23] With this may come deeper challenges to established assumptions, if the history of colonial science joins the fashionable literary company of 'subaltern studies' that seems intent upon 'provincialising Europe'.[24]

Among British historians of science, slow at first to read the literature of imperial history, this has imposed a sharp learning curve. It is not merely a question of distinguishing between white 'settler' colonies, with representative governments – that is, Europeans in 'European countries overseas', where indigenous populations were assimilated, segregated, or destroyed – and colonies of 'alien rule', where small numbers of colonists controlled settled indigenous populations, and who were governed directly from London or Paris, The Hague or Berlin. We also now have a much better picture of rival European imperial traditions, and competing philosophies of mercantilism and free trade.[25] For this reason, it is no longer easy, nor particularly desirable, to apply the same evolutionary language of institution-building to all nations and peoples once generally categorised together as the 'third world'. Indeed, 'tiers monde-ism' as an imperial development assumption invites refutation as a concept virtually irrelevant to the interpretation of local political meanings, interests and agencies.[26]

Just as definitions and categories and styles of imperialism have reflected broad differences in the colonial world, so have competing perspectives emerged in the historiography of colonial science. I have limited myself to a discussion of science within the British empire – a goal immodest enough, but one that proves less ambitious than the necessary comparison with contemporary experience in French, Dutch and German spheres of influence. In this context, attempts to categorise comparative differences between national styles in 'science overseas' have produced models of almost Ptolemaic complexity.[27] Deepak Kumar can easily be excused for speaking of a 'kaleidoscope of imperial science'.[28]

While such debate is healthy, constructive consensus and collective engagement with the issues has been confused by a cross-fire of approaches. Perhaps the single most long-running debate in the field – culminating in an inconclusive encounter in *Isis* in March 1993[29] – features, on the one hand, an historian who interrogates science as part of Europe's civilising mission through its extension overseas of the so-called 'exact' sciences. This argument is pursued in German, Dutch, and French colonial contexts, in which science appears as a 'vector of cultural imperialism'.[30] The exact sciences escape the contaminating influences of everyday political processes, which metropolitan scientists nevertheless manipulate for their own ends (chiefly, the ends of basic research). In the course of events, science conveniently supports the extension of metropolitan political power, a context in which colonial interests unavoidably descend to second place.

This view has produced much heat but little light. It proceeds from an extended syllogism, based on a fashionable but specious premise. Imperialism is an idea. Imperialism should shape science, but it does not, at least not the 'exact sciences'; therefore, science is not shaped by political ideas. Instead, science, at its most significant, is confirmed as a universalist, internalist, metropolitan, Enlightenment project – which of course individual scientists may, for their own political ends, manipulate as they can, but which escapes the slippery slopes of interpretative flexibility and social constructivism.

'Because the exact sciences resist ideological contamination, they may serve as a probe for studying how learning supports the extension of political power.'[31] This is a powerful statement. But, I suspect, as an orienting assumption, it no longer commands general acceptance. To privilege the exact sciences – for whatever reason internalist historians may find interesting – ignores much of what goes on in science generally; and much of what actually took place within the imperial programme. Worse, perhaps, it emphasises a centrifugal interpretation of science, and in so doing ignores the many centripetal tendencies returning knowledge from the colonial space. It focuses on civilising missions, and the assimilation of local cultures, without regard for colonial residents who seek to preserve those cultures. It is more concerned with agency than with contingency; with simple vectors, rather than with the ambivalences in which colonial life abounds. And of course, it rejects the importance attaching to indigenous knowledge practices and the symbolic meanings of natural knowledge, and in so doing, risks denigrating the 'knowledge spaces' of the locality, by privileging the knowledge spaces of the metropolis.[32]

This view we may – for argument's sake – call an essentialist position. From my point of view, however, it poses the wrong question. Science in an imperial context is not about ideological purity; it is about power, and like the language of orientalism, invites contextualisation. An alternative view – which does not, it seems to me, so much disagree with this position as distance itself from it – is offered by those historians who view an emphasis on the physicalist language of irradiation and transmission as misguided, and tending towards a monoculturalism that misconstrues the character of European encounters, and construes the colonial relationship as pure and simple, when in fact it was anything but. Such historians (and I count myself among them) prefer to see the extension of European science – just as art, literature, or music – as an inseparable element of European political and economic culture. Far from being uncontaminated, science serves, and is seen to serve, both imperial and colonial interests, and as much the interests of settlers and producers as of scientists themselves. There arise in the colonial world specialisms unknown to Europe, to which Europeans eventually pay tribute. Knowledge of plants and animals develops in ways shaped by the environment and a demand for interdisciplinary ways of knowing, while technologies take on adaptations that become, with time, both innovative and exportable. Colonial science is indeed influenced and shaped by the metropolis – but over the course of history emerge new models of the world, and new disciplines of research, previously unfamiliar in the metropole.

In some respects, this debate recalls the old internalist-externalist debate, but in a new guise – one that makes competing claims for the privileging of certain disciplines, and certain ways of thinking. Historians arriving fresh to this debate often comment on the fact that the scholars are apparently talking past each other – and are, moreover, overlooking the changing place held by science in the context of economic and technological, and also of cultural history. Inevitably, different theoretical assumptions favour different explanatory preferences. The debate is messy, and cries for agreement on boundary conditions. At the moment, closure seems elusive – resolution, distant. As historians have pursued different frames of meaning, it is not surprising that a general analytical framework has eluded us. Happily, however, a newer discourse is emerging in the meantime – one which promises to replace familiar, formulaic accounts of learned societies and government policies, with dynamic accounts of individuals and interests in action, focussing on the accidents and incidents of colonial life, and the facts of colonial discovery, appropriation and use. Science in this context becomes part of the history of colonial negotiation.

What is interesting and relevant about the current colonial discourse – flowing from the orientalist debate in literary scholarship – is its inherent appeal to a new universalism, which replaces a conception of European and western science as a Sartonian given, divisible at some definable point from oriental and traditional knowledge systems, with a new horizon along which the institutions of science respond to the colonial experience itself. In practical terms, this redefinition of categories requires historians to lift their eyes from government archives and official sentiments, and to look out the hermeneutic narratives of individuals, personalities and styles in research. A beginning has already been made in histories of trade, exhibitions and travel, which have informed readings of events to reveal the reciprocal nature of the colonial engagement.[33]

Today, GLOCAL (think globally, act locally) is a buzz-word of post-modernism. Some speak of a coming global ecumene that will diminish to vanishing point the asymmetries of colonialism. Others say that 'locality proclaims itself cosmopolitan', and fragments the global discourse; still others fear the disappearance of cultural differences under pressure from the scientific, commercial and technological superpowers.[34] The discipline of 'world history' has made a stunning comeback, partly in deference to shrinking humanities budgets and the niceties of political correctness, but also as a positive response to pressures for multicultural perspectives in western scholarship. As Homi Bhabha puts it, the culture of western modernity must be symbolically relocated from a post-colonial perspective.[35] The recently colonised demand to be considered not as ancient artefacts, but as modern artificers.

In this project, colonial historians and historians of colonial science are approaching a common task. Both must weigh how agency and contingency combine to transform the protocols of colonial power. It is clearly as impossible to view science in cultural isolation, as it is to envisage a view of nature that is ideologically neutral. European ideas have left an indelible mark upon post-colonial science. But the shortcomings of Europe's legacy have to be seen in the context of post-colonial expectations. Undoubtedly, British science in India helped to develop what Dipesh Chakrabarty has called a 'core culture' of imperialism. By the same token, it also created a 'culture-in-between-cultures', a component of colonising influence, but also a culture relevant to post-colonial identity. We need not dispose of our modernist belief in the internationalism of science, and the universalism of objective knowledge, to accept that rich and varied 'knowledge sites' not only exist and thrive throughout the world, many in cultures not our own, but also produce 'hybrids' essential to world science.[36]

A bridge between the historiography of colonial science and colonial discourse theory – as elaborated by Said, Bhabha, Thomas and others – seems to be necessary if we are to encourage a two-way traffic in science between donors and recipients, producers and clients. If we can move from a linear view of 'the expansion of Europe' towards a translational view of the interaction between Europe and overseas, we may see how the natural sciences, viewed as a product of European rationalism and

Enlightenment, were co-opted into a wider project of economic and political imperialism, leaving in their wake resentments and contradictions that remain part of the post-colonial dilemma.[37]

THE WAY FORWARD: CONSTRUCTING AN AGENDA

As a first step towards a new narrative of description and explanation, we look to find agreement on the leading characteristics and consequences of colonial science. If it is too early for a consensus – and there may in the end be none – there are at least some converging sentiments, and the rudiments of a common agenda. This agenda is conditioned by three concurrent developments in the historiography of the subject – a deeper knowledge of particular cases, a more careful use of comparative categories, and a selective appropriation of themes and concepts from interpretative discourses lying outside the traditional canon of the history of science – notably, anthropology, geography, and literary theory. At their intersection, we find elements of an approach which is solicitous towards agency and contingency, encouraging towards theory and description, and as cautious of motives as critical of outcomes.

The newer history of colonial science takes increasingly as its starting point the premise that all knowledge begins as local knowledge. Knowledge has, in the words of Shapin and Ophir, an 'irremediably local dimension', whether it is western knowledge transferred to a non-western site, or non-western knowledge imbricated in indigenous culture.[38] To paraphrase Foucault, knowledge is inscribed in place, and it is in the analysis of spaces, political and cultural, that we find the points of contact and lines of transmission that our study requires. In these heterogeneous spaces, the colonial world establishes 'counter-sites – heterotopias, as Foucault calls them – which mirror the experience of the metropolis and convey its sense of reality and meaning.[39] There is an international unity within which new knowledge is registered. But diversity, rather than uniformity, is a more obvious trait of nature, and dealing with diversity – environmental, anthropological, linguistic – is what colonial science was largely about.

In these knowledge sites, colonial science, in the structured form that emerges from the early nineteenth century, arrives as an extension of European nationalism, and matures within the lifetime of European imperialism. The institutions of science become imperial in themselves, and identified with the culturally isomorphic programme of nationalism and imperialism, authority and modernity. Science becomes – in different places, to varying degrees – a species of European currency. In their occupation of geographical space, Europeans establish claims to intellectual space. The world, defined and mapped by Europe, becomes a conceptual laboratory for the testing of European ideas of race, language, and culture.[40] Cultural possession, typically a product of military and economic conquest, is conceptually deepened by the naming of local strata, plants and animals. Colonised spaces support a vision – and a version – of the centre as exemplar.[41] Jermyn Street, Kew, Greenwich and the Jardin des Plantes become Latour's 'centres of calculation', staffed by brokers skilled at the arbitrage of ideas. These, both subject and object of political interest, use the quest for artefacts and their narrative description to legitimise the status of natural history and the pursuit of science.[42] As Foucault puts it, the natural sciences were not merely extensions of curiosity, but constituted a new domain of empiricity, at the same time describable and orderable.[43] Classification and order, reducing the unruly passions of nature to the ordered regularities of the library and the museum, the inventory and the balance sheet, become the hallmarks of the civilising mission. Conceptual unity is defined by learned communities, not by nature.

Inevitably, the metropolis is at first – and for decades – reluctant to let initiative pass to others, and seeks to define the terms of engagement. This remained the governing condition in the Dutch, German, Spanish and Japanese colonies, and in many French and former American possessions even today. In the eastern Portuguese empire, Goa, Timor and Macau canvass a 'devolved' model, run by locals,[44] while in Central and South America, creole interests turn European traditions into local ones. Within the British empire, different patterns arise in the colonies of North America, the Caribbean, southern Africa, India and the Indies, the Crown Colonies, the islands of the Pacific and Australasia, as well as Ireland.[45] In each place, locality imposed its own conditions. Within the settler colonies of the Empire, 'responsible government' eventually passes from the mother of parliaments to the Stormonts of science. The institutions of colonial science, typically central to the 'moving metropolis' – not unlike post offices, custodians of culture – structure what Benedict Anderson has called 'imagined communities', and in turn inspire incipient colonial nationalism.[46] By the middle of the nineteenth century, learned societies emerge as commodity exchanges, and colonial museums as banks, where local nature is audited and deposited in glass displays.[47] These institutions seal the connection, and create a new – and in this case, English-speaking, culture.[48] For a hundred years, colonial science, everywhere bred of European enterprise, becomes part of the civic humanism of colonial life. Then, however, gradually it finds its way into the struggle for nationhood. A tool of the coloniser, science would become an ally of decolonisation. From colonial America to the Raj in India, arguments for science that once underwrote imperial order and efficiency, were to be reinvented as arguments for national sovereignty.

By mid-century, the colonies of settlement, and those of alien rule, from Brazil to Bombay, Madras to Melbourne, Dublin to Dunedin, see scientists take up university and government positions, their own cultural *sodalitas* linking the counties of Ireland with the colonies of New South Wales and Victoria, the provinces of English Canada, and the Cape of Good Hope. In

North Africa and Indochina, the French establish what Michael Osborne has called 'tropical technocracies', administering metropolitan rule from peripheral centres. For both *savants* and gentlemen of science, to be a colonial was to know a kind of security, to inhabit a fixed world. In return for objects of virtue and intelligence, European institutions provided a service of referral, authority, and reward. Within that wider world, local elites emerge, linked to possession of local knowledge. Typically, colonial scientists underwent a process that Chakrabarty has discerned among Indians in British India, but which seems as widely evident among Europeans *outre-mer* – a 'double movement' of recognition, in which they took up uncertain positions between what appeared to be an ordered past and a less ordered present – in effect, a recognisably modernist condition. Although many took up residence, many remained in exile, from both their countries of origin and of choice. Psychological distance could not offset geographical proximity. Dublin, in the experience of Augustus De Morgan, was more 'central' than were his University College rooms in Bloomsbury.[49] The peculiar effects of distance, environment, and isolation produced remarkable strategies. Some became like 'K-competitors' in the animal world, seeking competitive niches in rich conceptual environments; others, like 'R-competitors', became resourceful 'edge dwellers' on the margins of professional habitation. Typically, the former left or returned 'home'; the 'edge dwellers' stayed, and taught, and begin new scientific traditions.[50]

As Europe dominates and rewards, research for most remains what Meaghan Morris calls the 'project of positive unoriginality'.[51] Divisions of labour are implanted between those in the metropolis who prepare the texts, and those at the periphery who inscribe them. Some trade in the criticism of European ideas, while others use local knowledge to register a place in the international discourse. Some occupy baronies in colonial astronomy, botany, and zoology. Others, more modestly, defend the bastions of public science. At first, theoretical leadership is remitted to the metropole, but in time, new as knowledge is remitted from the periphery, its ownership is claimed there. From the colonial perspective, the conduct of 'science overseas' then becomes a distinct activity in method, and choice of subject; its results are no longer 'transmitted' but 'exchanged', and its successes and discoveries, honoured at source. Its features, distanced from home, are accentuated by site and provenance. The world of science in the middle latitudes becomes an 'other' world, a laboratory of the outdoors; its methods, travel and exploration; its objectives, survey, collection, and display. Its motives, if once inspired by the Humboldtian quest to complete a picture of the world, eventually defines itself in local terms. Eventually, a post-colonial science develops, one not so much 'independent', in the language of Basalla, as 'co-dependent', sharing its dependencies with the metropolis and the rest of the scientific world. Its relationship with the European world becomes not linear, but dialectical, its traffic not one-way, but two. Within the colonies, mimicry and deference give way to translation. The imperial language of transplantation, irradiation, acclimatisation, absorption – western, physicalist and linear – becomes familial, organic, and reversible.

Long before the end of empire, the educated colonised grew bilingual, so to say, and in Joycean fashion, turned the Englishman's language against him. Today, historians are re-reading this language, and with it, the changing experience of colonial life. At present, this language is one of political correctness, in which exploration becomes domination, discovery becomes invasion, and enterprise becomes exploitation. The history of colonial science affords – even by name – many cases of 'experimental stations' whose research was of dubious benefit to indigenous inhabitants. Both India and Ireland were constructed, in W.L. Burn's classic phrase, as social laboratories for British policies.[52] Whether Irishmen or Indians saw themselves as experimental subjects possibly mattered less to them, than the image that this impression imparts. But the rules of engagement were severe on both coloniser and colonised. The responsibility of empire, as Lord Curzon put it, was 'equally . . . to dig and discover, to classify, reproduce and describe, to copy and decipher, and to cherish and conserve'.[53] That meant having points of reference, forms of calibration, which were external to places where new knowledge was found. In British India, it was hard duty, in Sir Henry Maine's phrase, 'to keep true time in two longitudes at once'. For better or worse, Britain rarely denied its colonies in peacetime any scientific knowledge it might possess itself; and by the same token, few colonial policies – and in this case, Britain was not alone – escaped the weakest features of their metropolitan models. Scientific research and technical education in Algeria or India, for example, may have languished for lack of government and industrial support, but such support was also hard to find in France or England. As in England, so in Ireland and India, local elites chose to neglect science, perhaps ensuring their momentary social survival, but also ensuring their absence from twentieth century science.[54]

In the new history of colonial science, historians are taking a polychromatic view of the world, in which the coloniser is no more equal than the colonised, and no more capable of independent alternatives.[55] Colonial science remained part of the colonial agenda. As such, it reflected separations between metropolitan and local cultures. Racialism, for example, never far from that agenda, became as explicit in doctrine as it was implicit in 'discovery'. The illusions of progress, similarly, came with the mental baggage. For Englishmen, as E.M. Forster once put it, India was not a promise, but an appeal. In India, Africa and Australasia as well, Europeans cast in evolutionary terms the differences between indigenous cultures and those that might be seen as part of a European past. It was easy, and necessary, to have a language of the progressive *versus* the non-progressive, the barbarous *versus* the civilised. That language entitled science to a place in the processes of orderly imperial improvement, from the pacification of the unruly savage, to the domestication of fertile Nature that 'ignorant savages' had failed to exploit. As the destruction of cultures preceded the discovery of their richness, so nature, once privileged as 'exotic', was plundered as an 'environmental resource'. As Michael Osborne has put it, this 'hundred years war against nature' has not yet come to an end.[56] Nor, at present, has the history of colonial science – its motives, methods and means inherited from Europe – yet achieved a balanced perspective in which such struggles – ones of place, space and time – are not dismissed as legacies of empire, but accepted as responsibilities of mankind.

CONCLUSION

At the convenient intersection of history, geography, and language, I have used several categories and concepts – locality, space, nationalism, division of labour, research strategies, modes of representation, and means of control – to suggest a framework for the discussion of colonialism, science, and the iteration of ideas between Europe and 'overseas'. Perhaps, as Michael Howard once put it, we need a readiness to think in analogies rather than theories, processes rather than structures, and to dwell on the politics of contingency. The categories we have used to describe colonial science, like the concepts of science, are historical constructs, which every generation must revisit and revise. Our mental models have changed. Thirty years ago, we were all diffusionists. Now, to be so seems impossibly simplistic. We have disposed of any notion that science in the colonial sense was *ipso facto* derivative. We have seen colonial models evolve into patterns of cooperation, no longer north-south, but increasingly south-south. And we see that even these markings on our compass rose are misleading and inaccurate, when we learn that even in Europe, the 'north' has a persisting 'south', and the 'south', or at least part of it, a promising 'north'.

Heidegger once said that boundaries are not that at which something stops, but that from which something else begins. During the last twenty years, we have become, perhaps inevitably, confined by our categories. Today, we see a world of colonial science using, by analogy, not one language called English, but many English languages. As the kaleidoscope turns and we learn more about different colonial 'pasts', we must not be so burdened with hindsight that we neglect the postcolonial experiment in which we live today. To paraphrase Sir William Harcourt, we are all postcolonials now. Given this fact, we cannot take up a new reading that fails to recognise its effects upon the indigenous world – specifically, upon peoples who, to paraphrase Marx, were thought unable to represent themselves, so for whom representations were made by others. In their envelopment by Europe, indigenous traditions and belief systems were reduced to artefacts, culturally inaudible except through their colonial interlocutors. Today, people insist on being heard, and in their own languages, no less in Ireland than in India. Surely this reading, and all it implies, must be listed as one of the central texts in the new history of colonial science, and in the study of its civilising passions.

NOTES

1. It is a pleasure to attend to the first conference devoted to science, technology and society in Ireland, and I wish to express my warm thanks to the organisers. I come from an island with a strong Irish tradition of its own, which no doubt helps account for its generous impulses – and if I had any temptation of 'sleeping through' this meeting, as a visiting Russian leader recently did some distance to the south, I could argue in my defence the peace and comfort of the accommodation afforded by my hosts. For travel expenses, I am indebted to the British Academy.
2. Oliver MacDonagh, *States of Mind: A Study of Anglo-Irish Conflict, 1780–1980* (London, 1983), p. 2.
3. This position has been developed with particular sophistication in India. See Deepak Kumar, 'Patterns of Colonial Science in India', *Indian Journal of History of Science*, 15 (1980): pp. 105–113, as contrasted with V.V. Krishna, 'The Colonial "Model" and the Emergence of National Science in India, 1876–1920', International Colloquium on Science and Empires (Paris: CNRS, 1990). See also Satpal Sangwan, 'The Sinking Ships: Colonial Policy and the Decline of Indian Shipping' and S. Irfan Habib, 'Science, Technical Education and Industrialisation: Contours of a Bhadralok Debate, 1890–1915', in Roy MacLeod and Deepak Kumar, eds., *Technology and the Raj: Technical Transfer and British India, 1780–1945* (New Delhi, 1995); and Shiv Visvanathan, *Organising For Science* (New Delhi, 1985), chapter 3.
4. Michael Adas, *Machines as the Measure of Men: Science, Technology and Ideologies of Western Dominance* (Ithaca, 1989).
5. MacDonagh, *States of Mind*, p. 6.
6. For Canada, see B. Sinclair, N.R. Ball and J.O. Petersen, eds., *Let us be Honest and Modest: Technology and Society in Canadian History* (Toronto, 1974) and Trevor H. Levere and Richard A. Jarrell, eds., *A Curious Field-book: Science and Society in Canadian History* (Toronto, 1974); for Australia, see Roy MacLeod, ed., *The Commonwealth of Science: ANZAAS and the Scientific Enterprise in Australasia, 1888–1988* (Melbourne, 1988) and Rod Home, ed., *Australian Science in the Making* (Sydney, 1988). For New Zealand, see M.E. Hoare and L.G. Bell, eds., 'In Search of New Zealand's Scientific Heritage', *Bulletin of the Royal Society of New Zealand*, No 21 (1984), and C.A. Fleming, 'Science, Settlers and Scholars', *Bulletin of the Royal Society of New Zealand*, No 25 (1987).
7. Edward Said's writings, beginning with *Orientalism: Western Conceptions of the Orient* (London, 1978) and continuing through *Culture and Imperialism* (London, 1993) have become the literary *locus classicus* of colonial discourse theory, to which more recently, Nicholas Thomas has made elegant contributions. See in particular Thomas, *Entangled Objects: Exchange, Material Culture and Colonialism in the Pacific* (Cambridge, Mass., 1991).
8. See Gili S. Drori, 'The Relationship between Science, Technology and the Economy in Lesser Developed Countries', *Social Studies of Science*, 23 (1993): pp. 201-15.
9. Anne Buttimer, 'Gatekeeping Geography through National Independence: Stories from Harvard and Dublin', *Erdkunde*, 49 (1995): pp. 1–16 and the same author's paper in this volume.

10. This has been recently exemplified in Roy MacLeod and Fritz Rehbock, eds., *Darwin's Laboratory: Evolutionary Theory and Natural History in the Laboratory of the Pacific* (Honolulu, 1994).
11. See Carlo Cipolla, *Guns and Sails in the Early Phase of European Expansion, 1400-1700* (London, 1965).
12. The details of this experience are now being extensively treated by a enterprising younger generation of scholars including Deepak Kumar, Satpal Sangwan, Michael Worboys, Michael Osborne, Christophe Bonneuil, Alberto Elena, and James McClellan. See Kumar, *Science and the Raj* (New Delhi, 1995); Bonneuil, 'Des Savants pour l'Empire' (Université de Paris VII, June, 1990); Worboys, 'British Colonial Science Policy, 1918–1939', in *20th Century Science: Beyond the Metropolis* (Paris: ORSTOM, 1993); Osborne, *Nature, the Exotic and the Science of French Colonialism* (Bloomington, 1994); and Elena: 'La Configuración de las periferias científicas: Latinoamérica y el mundo islámico,' in A. Lafuente *et al.*, Mundialización de la ciencia y cultura national (Madrid, 1993), pp. 139–146. See also McClellan, *Colonialism and Science: Saint Domingue in the Old Regime* (Baltimore, 1992).
13. This owed much to the fine work of, among others, Lewis Pyenson, Thomas Glick, Antonio Lafuente, Sverker Sorlin, and members of the history of science group in NISTADS, CSIR, New Delhi. See Pyenson's trilogy, *Cultural Imperialism and Exact Sciences: German Expansion Overseas, 1900–1930* (New York, 1985), *Empire of Reason: Exact Sciences in Indonesia, 1840–1940* (Leiden, 1989), and *Civilising Mission: Exact Sciences and French Overseas Expansion, 1830–1940* (Baltimore, 1993). For Glick, see 'Establishing Scientific Disciplines in Latin America: Genetics in Brazil, 1943–1960', in Lafuente *et al.*, *Mundialización de la ciencia y cultura nacional*, pp. 363–376.
14. Patrick Petitjean, Catherine Jami, and Anne-Marie Moulin, eds., *Science and Empires – Historical Studies* (Dordrecht, 1991).
15. George Basalla, 'The Spread of Western Science', *Science*, 156 (1967): pp. 611–622; Donald Fleming, 'Science in Australia, Canada and the United States: Some Comparative Remarks', *Proceedings of the Xth International Congress of the History of Science* (Ithaca, 1962), I: pp. 180–196, and I.B. Cohen, 'The New World as a Source of Science for Europe', *Actes du IX Congrés International d'Histoire des Sciences* (Madrid, 1960): pp. 96–130.
16. See, for example, Lucille H. Brockway, *Science and Colonial Expansion: The Role of the British Royal Botanical Gardens* (New York, 1979); James Secord, 'King of Siluria: Roderick Murchison and the Imperial Theme in Nineteenth Century British Geology', *Victorian Studies*, 25 (1982): pp. 413–442, and Robert A. Stafford, *Scientist of Empire: Sir Roderick Murchison: Scientific Exploration and Victorian Imperialism* (Cambridge, 1989).
17. Our understanding of the impact of tropical environments on European health, mortalities, and mentalities, owes much to Philip Curtin and, more recently, Richard Grove. See Curtin, 'The Environment beyond Europe and the European theory of Empire', *Journal of World History*, 1 (2), (1990): pp. 131–150, and Grove, *Green Imperialism* (Cambridge, 1995).
18. Jan Todd, 'Science at the Periphery: An Interpretation of Australian Scientific and Technological Dependency and Development prior to 1914', *Annals of Science*, 50 (1993): pp. 33–58.
19. Daniel R. Headrick, *The Tools of Empire: Technology and European Imperialism in the Nineteenth Century* (New York, 1981); *The Tentacles of Empire: Technology Transfer in the Age of Imperialism, 1850-1940* (New York, 1988); MacLeod and Kumar, eds., *Technology and the Raj*.
20. The language of core and periphery dates at least from its use in Wilbert E. Moore, 'Global Sociology: The World as a Singular System', *American Journal of*

Sociology, 71 (1966): pp. 475–482, and was taken up by Immanuel Wallerstein in *The Modern World-System* (New York, 1974), but probably owes its popularity to the wide reception accorded its use, in a different context, by Edward Shils, *Center and Periphery* (Chicago, 1975), chapter 1.
21. On this subject generally, see Lewis Pyenson 'Why Science May Serve Political Ends: Cultural Imperialism and the Mission to Civilize', *Berichte zur Wissenschaftgeschichte*, 13 (1990): pp. 69–81, and his survey, 'Science and Imperialism', in R.C. Olby *et al.*, eds., *Companion to the History of Modern Science* (London, 1990), pp. 920–933.
22. See Roy MacLeod, 'Museums in the Pacific: Reflections on an "Introduced Concept" in Transition', presented to the conference on '20th Century Science: Beyond the Metropolis' (Paris: ORSTOM, 1995).
23. A. Lafuente, A. Elena and M.L. Ortega, 'Un diálogo a tes bandas', in Lafuente *et al.*, *Mundialización de la ciencia y cultura nacional*, pp. 15–22.
24. Dipesh Chakrabarty, 'Postcoloniality and the Artifice of History: Who Speaks for the "Indian" Past', *Representations*, 37 (1992): p. 20. See also Gayatri Chakravorty Spivak, 'Subaltern Studies: Deconstructing Historiography', in Rajajit Guha and Gayatri Chakravorty Spivak, eds., *Selected Subaltern Studies* (Oxford, 1988), pp. 391–426.
25. Lafuente, *Mundialización de la ciencia y cultura nacional*.
26. See Edwige Lefebvre-Leclercq, *Tiers-Mondisme: Bridge Building and the Creation of the New Left in French Politics* (Cambridge, Mass., 1993) and Jean-Pierre Cot, *A l'épreuve du pouvoir: Le Tiers-Mondisme, pour quoi faire?* (Paris, 1984).
27. Among the most fascinating has been Togo Tsukahara's attempt to improve upon the three-dimensional analysis of imperial 'functionaries and seekers' proposed by Lewis Pyenson in 'Pure Learning and Political Economy: Science and European Expansion in the Age of Imperialism', in R.P.W. Visser *et al.*, *New Trends in the History of Science: Proceedings of a Conference held at the University of Utrecht* (Amsterdam, 1989), pp. 209–278; see Tsukahara, 'The Dutch Commitment to the Search for Asian Mineral Resources and the Introduction of Geological Sciences as its Consequence', presented to the Conference on Transfer of Science and Technology, Kyoto, November 1992.
28. Kumar, *Science and the Raj*, chapter 1.
29. Paolo Palladino and Michael Worboys, 'Science and Imperialism', *Isis*, 84 (1993): pp. 91–102.
30. Pyenson, *Civilising Mission*.
31. *Ibid.*, p. xiv.
32. For the colonial significance of local symbolic and ritual meanings, see David Arnold, 'Touching the Body: Perspectives on the Indian Plague', in Guha and Spivak (eds.), *Selected Subaltern Studies*, pp. 391–426; and David Turnbull, 'Moving Local Knowledge', paper presented to the Ford Foundation Conference on 'Understanding the Natural World: Science Cross-Culturally Considered', Amherst, June, 1991. In this respect, I find the work of Sandra Harding is particularly illuminating. See her 'Eurocentric Scientific Illiteracy: A Challenge for the World Community', in Harding, ed., *The 'Racial' Economy of Science: Toward a Democratic Future*, (Bloomington, 1995).
33. See Nicholas Thomas, *Colonialism's Culture: Anthropology, Travel and Government* (Melbourne, 1994).
34. Ulf Hannerz, 'Notes on the Global Ecumene', *Public Culture*, 1 (2), (1989): pp. 66–75; See 'World Science: Globalization of Institutions and Participation', *Science, Technology and Human Values*, 18 (1993): pp. 196–208.
35. The new language of hybridity, and the task of 'destabilising binaries' – filling in 'spaces' between, for example, the European and the 'other' – forms part of

this fashionable discourse. See Homi K. Bhabha, *The Location of Culture* (London, 1994).
36. Michael Chayut, 'The Hybridisation of Scientific Roles and Ideas in the Context of Centres and Peripheries', *Minerva*, XXXII (1994): pp. 297–308.
37. An attempt from a metropolitan perspective is proposed in Roy MacLeod, 'Passages in British Imperial Science: From Empire to Commonwealth', *Journal of World History*, 4 (1993): pp. 2–29.
38. Adi Ophir and Steven Shapin, 'The Place of Knowledge: A Methodological Survey', *Science in Context*, 4 (1991): pp. 3–21.
39. Michel Foucault, 'Of Other Spaces', *Diacritics*, 16 (1986): p. 24.
40. Roy MacLeod and P.F. Rehbock, eds., *Darwin's Laboratory: Evolutionary Theory and Natural History in the Pacific* (Honolulu, 1994), Introduction.
41. See Dhruv Raina, 'Epistemic Drift between Centre and Periphery: The Structure of Scientific Exchanges in Colonial India', presented at '20th Century Science: Beyond the Metropolis' (Paris: ORSTOM, 1994).
42. See Nicholas Thomas, 'Licensed Curiosity: Cook's Pacific Voyages', in John Elsner and Roger Cardinal (eds.), *The Cultures of Collecting* (Melbourne, 1994).
43. Michel Foucault, *The Order of Things: An Archaeology of the Human Sciences* (1966, rep. New York, 1973), ch. 5 'Classifying', p. 158.
44. I am indebted for this observation to Dr Teresa Patricio, and to her use of Portuguese materials in Lisbon.
45. See, e.g., Richard Jarrell, 'Differential National Development and Science in the Nineteenth Century: The Problems of Quebec and Ireland', in Nathan Reingold and Mark Rothenburg, eds., *Scientific Colonialism: A Cross-Cultural Comparison* (Washington, DC, 1987), pp. 323–350.
46. Benedict Anderson, *Imagined Communities: Reflections on the Origin and Spread of Nationalism* (London, 1983).
47. Susan Sheets-Pyenson, *Cathedrals of Science. The Development of Colonial Natural History Museums during the Late Nineteenth Century* (Kingston, Ont., 1988), and Sally Gregory Kohlstedt, 'Australian Museums of Natural History: Public Priorities and Scientific Initiative in the Nineteenth Century', *Historical Records of Australian Science*, 5 (1983): pp. 1–29.
48. Roy MacLeod, ed., *The Commonwealth of Science*.
49. Ian Inkster, 'Scientific Enterprise and the Colonial "Model": Observations on Australian Experience in a Historical Context', *Social Studies of Science*, 15 (1985): pp. 677–704.
50. Here I acknowledge the theoretical insights of Stephen Budiansky, *The Covenant of the Wild* (New York, 1992). On contrasting research styles, see MacLeod, *Commonwealth of Science*, chapter 1.
51. Meaghan Morris, 'Metamorphoses at Sydney Tower', *New Formations*, 11 (1990), p. 10, as cited in Chakrabarty, 'Postcoloniality and the Artifice of History', p. 17.
52. W.L. Burn, 'Free Trade in Land: An Aspect of the Irish Question', *Transactions of the Royal Historical Society*, 4th ser, 31 (1949), p. 68. For a critical discussion of the 'social laboratory' model as applied to colonial science, see Mark Harrison, *Public Health in British India: Anglo-Indian Preventive Medicine, 1858–1914* (Cambridge, 1994).
53. Quoted in Anderson, *Imagined Communities*, p. 179.
54. See Stephen Yearley, 'Colonial Science and Dependent Development: The Case of the Irish Experience', *The Sociological Review*, 37 (1989): pp. 308–331, and several essays by Kapil Raj on India.
55. MacLeod and Kumar, eds., *Technology and the Raj*.
56. Osborne, *Nature, the Exotic, and the Science of French Colonialism*.

Chapter 2

The Social Contexts of W.R. Hamilton's Prediction of Conical Refraction

David Attis

INTRODUCTION[1]

On the 22nd of October, 1832, William Rowan Hamilton, Royal Astronomer of Ireland and Professor of Astronomy at Trinity College Dublin, announced the mathematical discovery of a novel optical phenomenon to a meeting of the Royal Irish Academy. Using Fresnel's wave theory of light and the sophisticated mathematical methods that he had developed in his previous studies of geometrical optics, Hamilton predicted that a single ray of light could, under the proper circumstances, be refracted into a cone of light within a biaxial crystal. It was a stunning prediction based entirely on mathematical theory and without physical precedent. When experimentally confirmed by Hamilton's Dublin colleague, Humphrey Lloyd, it created a sensation in the world of nineteenth century British physical science.[2]

Reaction was intense. William Whewell of Cambridge explained in his opening address to the British Association for the Advancement of Science (BAAS) the following year, 'In the way of such prophecies, few things have been more remarkable than the prediction [of conical refraction].'[3] George Airy, another Cambridge mathematician, exclaimed, 'Perhaps the most remarkable prediction that has ever been made is that lately made by Professor Hamilton.'[4] Hamilton immediately became a superstar of the British scientific world. In 1835 he was knighted by the Lord Lieutenant at the meeting of the BAAS in Dublin and was awarded the Royal Medal of the Royal Society for 'discoveries in Optics, and particularly that of Conical Refraction'.[5] Even after the immediate furore, conical refraction continued to be seen as a miraculous prediction. Charles Babbage used it in his *Ninth Bridgewater Treatise* as evidence for his argument that what appears to be miraculous is actually governed by scientific laws,[6] and six years after the discovery, the German mathematician Julius Plücker wrote 'Aucune expérience de physique a fait tant d'impression sur mon esprit, que celle de la réfraction conique.'[7] Even recent commentators present it as a paradigm of scientific prediction.[8]

Although there are a number of excellent accounts of Hamilton's prediction, all of them focus on the context of mathematical optics: the history of the theory of double refraction, the specific ways in which Hamilton deduced the prediction from Fresnel's wave theory, Lloyd's experimental set-up, etc.[9] Such a narrow approach makes it difficult to account for two aspects of the prediction. First of all, it makes it difficult to understand the excitement it generated. As O'Hara observes, 'Conical refraction is little more than a curious optical phenomenon which had no conceivable application.'[10] And Hankins explains, 'It is a curiosity, comparable to many other optical phenomena discovered in the early nineteenth century...'[11] Hamilton's prediction added nothing to the physical principles of the wave theory of light, nor, as Stokes later pointed out, did it necessarily prove Fresnel's theory since other wave surfaces could be imagined with the same physical effect.[12] It does not even seem to have converted any of the proponents of the corpuscular theory or changed the positions of the Cambridge mathematicians who were already convinced of the truth of the wave theory.[13]

Previous accounts also gloss over the question of why Hamilton chose to investigate Fresnel's wave theory. Since his previous research was in optics, they take it for granted that he would be interested in Fresnel's wave theory, making conical refraction appear to be the inevitable outcome of his earlier research. Sarton even refers to Hamilton's earlier work as 'a series of papers wherein the wave theory of light, then recently elaborated by AUGUSTIN FRESNEL, was brilliantly developed',[14] when in fact Hamilton's earlier papers were on geometrical optics and explicitly avoided the question of the physical nature of light.[15]

Although Hamilton developed mathematical methods in his earlier work that were well suited to the investigation of the Fresnel wave surface, this alone does not explain why Hamilton would choose to work on the wave theory of light. In fact, Hamilton wrote to the poet Coleridge just after his prediction, explaining that in optics, 'My aim has been, not to discover new phenomena [but] . . . to remould the Geometry of Light . . .'[16] and he went on to describe the discovery of conical refraction as 'only a secondary result, my chief desire and direct aim being to introduce harmony and unity into the contemplations and reasonings of Optics, considered as a portion of pure Science'. Thus although conical refraction was well-received, it was in some sense contrary (or at least secondary) to Hamilton's goals in optics. This is clear in a letter he wrote to John Herschel immediately after the discovery, 'You are aware that the fundamental principle of my optical methods does not essentially require the adoption of either of the two great theories of light in preference to the other. However I naturally feel an interest in applying my general methods to Fresnel's theory of biaxial crystal . . .'[17] Hamilton's interest, I would argue, and the interest shown by Hamilton's contemporaries in what was simply 'a curious optical phenomenon' cannot be understood simply in the narrow context of mathematical optics.

Other accounts of conical refraction present it as a paradigm of good science – a risky theoretical prediction confirmed by quantitative experiment – and they take the value of this style of science to be self-evident. However, I will show that at the time of the prediction those very standards were being debated by scientists and non-scientists alike. Not only was the question of whether light is a wave or a particle hotly debated, but disputants on either side invoked differing methodologies to support each theory. These methodological debates were also linked to broader questions about the proper type of science and the proper role for science in British culture. Additionally, many political and social arguments in mid-nineteenth century Britain looked to science as the standard of objective and true knowledge, and for this reason, arguments in the philosophy of science could have important repercussions in many other areas of debate. By looking at a number of separate but related contexts – Section A of the BAAS, the debate between the idealists and the utilitarians, Trinity College Dublin and the Protestant Ascendancy of Ireland – I hope to show some of the many ways in which Hamilton's prediction of conical refraction was both shaped by its contexts and subsequently served as a means by which Hamilton and others attempted to shape those contexts according to their own goals.

Waves vs. particles at the BAAS

The primary arena for British physical science in the 1830s was the British Association for the Advancement of Science. Many of the most important events in the debate over the physical nature of light occurred there, and so in some sense it is the most immediate context for conical refraction. Within the BAAS were a number of factions, all vying for certain scientific, political and religious goals, and as a number of authors have suggested, the debate over the nature of light involved many of these aspects.[18] Conical refraction was an important event in that debate, and though it did not prove Fresnel's wave theory or end the debate over the physical nature of light, it represented the triumph of a certain style of science over another opposing style of science and of the faction of Cambridge and Dublin mathematicians who championed that style over the (primarily) Scottish experimentalists who opposed it.

For the wave theory's opponents, Henry Brougham, David Brewster and Richard Potter, the theory involved unwarranted hypotheses. Brougham and Brewster had been educated in Edinburgh where

Common Sense philosophy was the dominant ideology. They were a part of what Cantor has called the Scottish methodological tradition, a tradition that took as its starting point the writings of Bacon, Newton and Locke and was exemplified in the work of Thomas Reid, Dugald Stewart, John Robison, John Playfair, John Leslie and Joseph Black.[19] Following Newton's claim 'hypotheses non fingo', the Scots were opposed to hypotheses and advocated (following Bacon) the use of gradual generalisations from experimental evidence. They sharply distinguished between facts which they believed could be perceived immediately by the mind and theories or hypotheses which they believed to be simply guesses or useful fictions.

While many of the Scots believed that hypotheses could play a role in guiding experiment, it was experimental facts that were the foundation and goal of all science. John Robison, an Edinburgh natural philosopher and one of Brewster's teachers explained,

> a fancied or hypothetical phenomenon can produce nothing but a fanciful cause, and can make no addition to our knowledge of real nature . . . Although all the legitimate consequences of a hypothetical principle should be perfectly similar to the phenomenon it is extremely dangerous to assume this principle is the real cause.[20]

Robison doubted that any hypothesis could be proven true by any amount of evidence. Brewster himself explained, 'We have no objection to hypotheses, however wild, when they are used but as incentives or as guides to observation and experiment, but we reject them with disdain, whether they are brought forward as true themselves, or as the ornaments or bulwarks of truth.'[21] Brewster admired the wave theory and even accepted the law of interference as an experimental fact, but he did not believe the wave hypothesis to be true. He felt that the proponents of the wave theory made dogmatic claims about its veracity and that such claims were dangerous to the progress of science.

The mathematicians of Cambridge and Dublin, on the other hand, supported the wave theory with their own opposing methodological and epistemological arguments. For them it represented an ideal of a mathematical, predictive science in which abstract theories are confirmed by experiment. Whewell attacked the opposing view in his speech to the BAAS in 1833 (the same speech in which he praised the prediction of conical refraction),

> It has of late been common to assert that *facts* alone are valuable in science; that theory, so far as it is valuable, is contained in the facts; and, so far as it is not contained in the facts, can merely mislead and preoccupy men . . . But . . . it is only through some view or other of the *connexion* and *relation* of facts, that we know what circumstances we ought to notice and record . . .[22]

Whewell later expanded this view of science into a complete philosophy in his *Philosophy of the Inductive Sciences*.[23] In that work he argued that all knowledge (even the most simple observation) is a combination of fact and theory, and he attempted to outline the ways in which we can be confident that our theoretical knowledge is true or approaching the truth. Hamilton's philosophy of science similarly emphasised the importance of ideas and theory over bare collections of facts.[24] His goal was, 'the exhibition of *a deductive rather than an inductive unity* in our contemplation and knowledge of nature, a Kantian rather than a Baconian connexion between the several parts of physical science'.[25]

For Whewell and Hamilton as well as other Cambridge and Dublin mathematicians such as John Herschel, George Airy and Humphrey Lloyd a confirmed prediction like conical refraction provided important evidence for the truth of the wave theory, whereas for Brewster and the other corpuscularians, no amount of evidence could prove a theory. In fact, in response to Hamilton's successful prediction Brewster wrote,

> I have long been an admirer of the singular power of this [wave] theory to explain some of the most perplexing phaenomena of optics; and the recent beautiful discoveries of Professor Airy, Mr. Hamilton, and Mr. Lloyd afford the finest examples of its influence in predicting new phaenomena. The power of a theory, however, to explain and predict facts, is by no means a test of its truth . . .[26]

Brewster's criticism was not of the veracity of the prediction but of its implications and underlying philosophy. Conical refraction was continually referred to as a prediction without precedent or analogy, in other words, a prediction that could never have been made using the empirical methods of the proponents of the corpuscular theory who stressed the importance of analogy in generalising from experiments. Thus the successful prediction of conical refraction was seen as a triumph of mathematical and theoretical approaches to science over simple empiricism. Augustus De Morgan explained that he believed the discovery of conical refraction to be 'a most important one, as a *predicted* result, in the very teeth of all former experience . . . important to the *philosophy* of induction'.[27] It is for this reason that the discovery of a curious phenomenon with no practical application was felt to have great scientific importance.

IDEALISTS VS. UTILITARIANS IN BRITAIN

In addition to their methodological and philosophical differences, the proponents of the wave and particle theories of light also differed on social and political issues. Brewster and Brougham were both Edinburgh educated, they made their livings outside of universities, they were Dissenters and reform Whigs (Brougham, in fact, was an MP and Lord Chancellor). They both opposed all that the English universities stood for. Meanwhile the proponents of the wave theory were almost all professors at Cambridge, Oxford and Trinity College, Dublin. They have been characterised by Morrell and Thackray as supporters of the 'Cambridge Programme'. They explain,

> In the 1830s Section A of the British Association became the familiar haunt of a group of Anglican, mainly clerical, gentlemen committed to the placement of physical science within the dominion of mathematical analysis. What may conveniently be called the Cambridge programme had implications of a moral, institutional and career nature, in addition to mathematical and practical aspects. Subscribing to or opposing that programme was not simply a matter of how one interpreted certain experiments in optics, though that was one central question. It was also a matter of commitment to, or rejection of, the growing role of the English universities in British life. And, not surprisingly, commitments of this kind were closely interwoven with wider religious and philosophical views.[28]

For this reason the important methodological implications of conical refraction could be (and were) tied to other institutional, religious, social and political goals.

Even outside the context of the BAAS, however, conical refraction was seen to have important implications. In response to the prediction, Hamilton's Irish friend Aubrey De Vere wrote, 'this sort of *á priori* science seems to me its utmost and ultimate triumph. I confess I like to see experiment occasionally put to the wheel, and reason harnessed as leader in these utilitarian times.'[29] Hamilton and De Vere often discussed their fear of utilitarianism and felt that only metaphysics and idealism could save the nation from its destructive philosophy. De Vere summarised his fears in relating to Hamilton an incident in which he confronted a number of utilitarians,

> The three gentlefolks differed in some respects but agreed in these enlightened principles of modern philosophy: – 'there is no natural, necessary, or eternal right or wrong; our impressions of those subjects are only associations instilled in us during childhood, for the good of society; the human mind has no natural *principles* of beauty, much less *Idea* of beauty; there is no such thing as conscience; morality is a mere name . . . the only true method of pursuing metaphysical subjects is experience; and Bacon's Inductive philosophy is the key to all philosophy; the first desire of every man is and ought to be his own happiness . . .' These doctrines are, I am afraid, terribly prevalent these days: and if so, what hope is to be entertained for a nation consisting of men who believe them?[30]

De Vere linked associationist psychology, Baconian empiricism and utilitarian ethics and felt that their rise represented the moral and philosophical decline of the nation. Nor was De Vere alone in this belief. Hamilton's friends William Whewell and William Wordsworth as well as his intellectual idol Coleridge expressed similar beliefs.[31] De Vere ended his letter to Hamilton, 'How is Coleridge's health now? Is he at the Logos? I am afraid even that book will not be able to stem the torrent of corruption that is flooding the country.' They all feared for the state of the nation and felt that the only solution lay in philosophy, particularly the philosophy of science. Science as the standard of knowledge was utilised in debates about philosophy, ethics and political theory. Because of its importance to the philosophy of science, therefore, the prediction of conical refraction was seen by some to have a social significance.

The debate between idealists and utilitarians was much broader than Hamilton's small circle of friends. In an 1840 essay, J.S. Mill even claimed that 'every Englishman of the present day is by implication either a Benthamite or a Coleridgian',[32] and he proceeded to analyse this dichotomy in terms not only of epistemology but also ethics and political theory. For disputants on both sides theories of knowledge were seen as entailing a certain social order.[33] The Benthamites adopted a Lockean empiricist epistemology based on the notion that all knowledge is generalisation from experience.[34] Mill explains that Bentham believed that 'abstractions are not realities *per se*, but an abridged mode of expressing facts'.[35] For him sensation is the sole source of knowledge, and metaphysics is simply mysticism. Ideas are only copies of sensations.[36] Bentham also used this philosophy of knowledge to develop the utilitarian system of ethics, arguing that since there is no such thing as an inherent moral sense, all ethical decisions must be based on empirical facts: in simplistic terms, the greatest good for the greatest number.

Utilitarianism also held that like abstractions, political institutions are simply artificial contrivances of convenience. They are nothing more than the sum of their parts and so should reflect the wishes of the majority of their components. Mill explicitly linked this philosophy to reform,

> The practical reformer has continually to demand that changes be made in things which are supported by powerful and widely-spread feeling, or to question the apparent necessity and indefeasibleness of established facts; and it is often an indispensable part of his argument to show, how those powerful feelings had their origin, and how those facts came to seem necessary and indefeasible. There is therefore a natural hostility between him and a philosophy which discourages the explanation of feelings and moral facts by circumstances and association, and prefers to treat them as ultimate elements of human nature . . .[37]

While the proponents of the corpuscular theory cannot all be classified as Benthamites, their theories of knowledge and society have close parallels (James Mill, in fact, was a classmate of Brewster and Brougham in Edinburgh). Hamilton's and Whewell's (though not necessarily all wave theorists') fight against the corpuscular theory and its Scottish proponents can be seen as separate though closely related to their battle against the utilitarians.

For idealists such as Coleridge, Wordsworth, Hamilton and Whewell, knowledge is impossible without ideas which they believed to be a precondition for experience. But idealist epistemology also opposed Benthamism on ethical and political issues. Fundamental ideas, they argued, are the basis of ethics just as much as they are the basis of mathematics.[38] Similarly, idealists saw political institutions as based on fundamental ideas and therefore not subject to the desires of the majority.[39] All of these idealists were conservatives, and they enlisted all of the above arguments in their fight against the reforming Benthamites.

The battle between Mill and Whewell is already well-known.[40] Whewell devoted his life to defining proper science through sermons, moral tracts, science textbooks and his *History* and *Philosophy*. Just as he had in his 1833 speech to the BAAS he argued that facts are meaningless without theories and that even simple perceptions require inference by an active mind. The implication of this belief is that knowledge is possible only through the work of theorists, who are primarily professors at British universities, and Mill observed that the tendency of Whewell's efforts, 'is to shape the whole of philosophy, physical as well as moral, into a form adapted to serve as a support and a justification to any opinions which happen to be established'.[41] Mill wrote his *System of Logic* explicitly in opposition to Whewell's philosophy as an empirical antidote to idealism whose political implications he feared.

Wordsworth and Coleridge also played crucial roles in developing the idealist position. Romantics are often portrayed as anti-scientific, but Wordsworth explained his beliefs to Hamilton when he visited Ireland in 1829, defending himself,

> from the accusation of any want of reverence for Science, in the proper sense of the word – Science, that raised the mind to the contemplation of God in works, and which was pursued with that end as its primary and great object; but as for all other science, all science which put this end out of view, all science which was a bare collection of facts for their own sake, or to be applied merely to the material uses of life, he thought it degraded instead of raising the species . . . and what is disseminated in the present day under the title of 'useful knowledge', being disconnected, as he thought it, with God and everything but itself, was of a dangerous and debasing tendency.[42]

Romantics like Wordsworth and Coleridge were not opposed to science but to a certain kind of science: useful knowledge. This was not simply a vague fear of practical science. The Society for the Diffusion of Useful Knowledge (SDUK) was a programme headed by Brougham and the utilitarians who saw such education as part of a larger programme of social reform. Beginning in the 1820s the SDUK and the Mechanics Institutes attempted to educate and empower the working class through practical education.[43] Noted conservatives (and Romantics) Wordsworth, Coleridge and Southey argued that the diffusion of useful knowledge would lead to superficial knowledge and loss of respect for profound thought (as well as for authority). Coleridge argued, 'You begin . . . with the attempt to *popularize* science: but you will only effect its *plebification*.'[44] They opposed bare empirical knowledge on political and moral grounds.

In this battle between utilitarianism and idealism, mathematics and poetry (Hamilton's two great loves) played a crucial role for they were extremely difficult for empiricists to account for. An epistemology based solely on sense perception found it almost impossible to explain artistic creativity (Bentham is said to have claimed that 'All poetry is misrepresentation') or mathematical truth (though Mill tried). Conical refraction was a particularly good example of a mathematical prediction that could not be explained on the view that mathematics is simply generalisation from experience. It could be seen as proof of the priority of ideas and therefore as proof of the entire idealist agenda. Mill recognised the danger when idealism used science as its support in this way. He explained,

> The notion that truths external to the mind may be known by intuition or consciousness, independently of observation or experience, is, I am persuaded, in these times, the great intellectual support of false doctrines and bad institutions ... And the chief strength of this false philosophy in morals, politics and religion, lies in the appeal which it is accustomed to make to the evidence of mathematics and of the cognate branches of physical science.[45]

Like Whewell and Coleridge, Hamilton used idealism to oppose reform but rather than describing the best kind of science as his friends did, he produced an example of it.[46]

Ascendancy vs. Democracy in Ireland

While it should be clear by now why others – members of the Cambridge programme and idealists – were interested in Hamilton's prediction, it is still not clear why Hamilton decided to investigate Fresnel's wave theory in the first place. Though I have portrayed Hamilton as an important member of the Cambridge programme at the BAAS, he differed from his Cambridge colleagues in many ways and his choice to work on Fresnel's theory was the result of a compromise, a compromise designed to solidify his relationship to the Cambridge mathematicians and to make a name for himself in British science. This decision can only be understood in terms of Hamilton's membership in another group, the Protestant Ascendancy in Ireland, and in terms of the relationship of Ascendancy culture to English culture.

Hamilton's first contact with the Cambridge mathematicians had occurred in 1827 when he competed with Airy for the post of Royal Astronomer of Ireland. Airy explained to him that there was little contact between Cambridge and Trinity College Dublin and that he hoped to change that situation. Closer contacts between the two universities would have been desirable on both sides. They both shared a certain view of the role of British universities in the life of the country; they were members (many of them clergymen) in the Anglican church, and they taught similar types of mathematics.[47]

However, while they shared many of the same goals, Hamilton's work was still very different from that of his Cambridge colleagues. He once explained to De Vere, 'I differ from my great contemporaries, my "brother-band", not in transient or accidental, but in essential and permanent things: in the whole spirit and view with which I study Science.'[48] While they were interested in understanding the physical nature of light, Hamilton was looking for a general theory of geometrical optics which was interpretable in terms of either the wave or particle theory of light. He was a devotee of Kant and Coleridge and more idealistic in his philosophy even than Whewell. Even with the evidence provided by conical refraction he was often hesitant to call the wave theory true and sought instead for truths in mathematical dynamics and algebra that were above empirical confirmation. He was more interested in *a priori* truth than empirical truth while Cambridge mathematicians like Whewell sought a synthesis of the two. Hamilton also differed with his Cambridge colleagues on the foundations of algebra.[49]

However, when the main goals of the Cambridge programme were challenged by the proponents of the corpuscular theory, Hamilton's interests were on the side of the Cambridge men. Also, as an Irishman, it was important for Hamilton to gain the respect of his English colleagues, and in order to gain the respect of the Cambridge mathematicians Hamilton had to compete on their terms. While no one would deny the skill with which he had written about geometrical optics, no one was really interested in it. In England and Scotland the question of the physical nature of light was paramount, and Cambridge mathematicians came down firmly on the side of Fresnel's wave theory.

Hamilton's compromise, I would argue, must therefore be seen in the context of Anglo-Irish relations, for one aspect of Hamilton's decision was a desire to prove the Irish capable of great scientific accomplishments. Irish Protestants for the most part ignored the culture of the native Irish and sought their culture instead in England.[50] They felt their ascendancy over the Catholics was cultural as well as political, and the main organ of this cultural and intellectual ascendancy was Trinity College Dublin which had been established in the fifteenth century as an outpost of the Anglican church and English culture.[51] Its main goal was to train men to become members of the establishment, and its Fellows were clergymen and major landholders. Their entire way of life depended on the Protestant Ascendancy in Ireland, and their intellectual role was to support that ascendancy by keeping English culture and Anglican theology alive in Ireland.

The relationship of the Ascendancy to England, however, was not simply one of adulation. Foster has characterised it as one of 'conscious but resented dependence'[52] also represented in Irish political institutions, and science played an important part in that relationship. The rise of science in late eighteenth century Ireland was contemporaneous with Irish gains in independence from England. In 1782 the Declaratory Act was repealed and Poynings' law amended, giving the Irish parliament a new-found independence and leading Grattan to declare the birth of the Irish nation.[53]

Soon after, in 1785, the Royal Irish Academy was founded.[54] Another element of this cultural movement was the establishment of the observatories of Dunsink and Armagh. With the help of Nevil Maskelyne, Astronomer Royal for England, and Jesse Ramsden, the famous London instrument maker, Dunsink Observatory was opened in 1785.[55] Armagh Observatory was completed in 1790, and its first astronomer was also in close contact with Maskelyne.[56] Science, therefore, represented both closer links to England and an assertion of the abilities of the Irish.

With the coming of the Union in 1801, however, Ireland was once again subordinated to England. Members of the Ascendancy gave up a measure of their independence for the security England could provide them.[57] With the removal of many of Dublin's best and brightest to London, however, science suffered. Hamilton's appointment as Royal Astronomer should be seen in this context. It was designed to improve the reputation of Trinity College, known since the eighteenth century as the 'Silent Sister' because of its relative lack of scholarly contributions. Though without experience in practical astronomy, Hamilton was their best hope for producing new research in the French analytical style of mathematics which had recently gained popularity. He later wrote, 'I should like to contribute my mite, or shall I say, my stone to throw upon the pile which hides the buried slander against the "Silent Sister" . . .'[58] He realised from the beginning that his research was important not only for his own fame but also for that of his university and his country. This helps explain why Hamilton would have been interested in pursuing research of interest to the mathematicians at Cambridge, for they were the primary audience for Hamilton's science. Thus by courting their favour Hamilton could increase the reputation of Trinity and the Irish in general. De Vere recognised this and wrote to Hamilton just after his prediction of conical refraction, 'I can most entirely sympathise with the exultation you must feel at the success of your mathematical discovery. I should think from its connexion with Physics, the popular part of Science, it is more likely to enlarge the "crescent sphere" of your fame than anything else you have done . . .'[59]

Hamilton believed that increasing the reputation of Ireland in this way could have very concrete benefits for daily life in Ireland. When De Vere later wrote to him during the height of the famine to ask what he was doing to help, Hamilton responded,

> It is the opinion of some judicious friends . . . that my peculiar path, and best hope of being useful to Ireland, are to be found in the pursuit of those abstract and seemingly unpractical contemplations to which my nature has a strong bent. If the fame of our country shall be in any degree raised thereby, and if the industry of a particular kind thus shown shall tend to remove the prejudice which supposes Irishmen to be incapable of perseverance, some step, however slight, may be thereby made towards the establishment of an intellectual confidence which cannot be, in the long run, unproductive of temporal and material benefits . . .[60]

By proving to the English that the Irish are capable of great accomplishments Hamilton (and his friends) believed his work to be crucial to the well-being of Ireland. Thus science in Ireland (almost the exclusive preserve of the Protestant Ascendancy) was an important part of the cultural and political relationship with England. There was a belief among the educated Irish that scientific achievement could strengthen both the Union and the position of Ireland within the Union. The BAAS itself was clearly a symbolic representation of the union which science could achieve and Hamilton's speeches to the BAAS often reflected on this.[61]

Hamilton's science could also serve as a support for the Ascendancy in the same way it was seen to support the Establishment in England – by representing Protestants as the sole protectors of truth. Since Daniel O'Connell and some fellow-barristers had formed the Catholic Association in 1823, the position of the Ascendancy in Ireland was increasingly jeopardised.[62] Never before had the Catholics been an organised force for social and political change. O'Connell also declared allegiance to the theories of Bentham and even corresponded with him for a time. Thus O'Connell represented all that Hamilton feared: utilitarianism, democracy and the end of the Ascendancy. Hamilton feared that if the educated and cultured Protestant Ascendancy were no longer allowed to govern then all society would crumble, and during the 1830s this did not seem like an impossibility. Hamilton himself joined the Protestant Conservative Society in 1834 and at his only public political appearance claimed that 'a design is evidently entertained, and covered by only the flimsiest veil, which often scarcely mocks us with concealment, to establish the Roman Catholic church in Ireland upon the utter ruin of the Protestant religion'.[63] Hamilton was extremely concerned about the political changes in Ireland, and I have attempted to argue that he (and others) saw his science as a valuable way to fight those changes and preserve all that he believed to be right in Ireland.

CONCLUSION

In his centennial essay George Sarton explains,

> Some sceptical men of science like to deny the absolute validity of scientific theories, and to consider them only as clever means of accounting for the facts . . . However such objections are untenable with regard to predictions like those made by HAMILTON, and a little later by ADAMS and LEVERRIER, for in such cases the facts follow the theories.[64]

For Sarton as for Hamilton and many of his contemporaries the primary importance of conical refraction was as an illustration of what he believed to be the proper philosophy of science. By situating it only within the history of mathematical optics other accounts tacitly assent to Sarton's philosophical claims; they make it appear self-evident that Hamilton would have devoted his time to such a study and that other scientists would have been interested and excited by such a prediction. I have tried to demonstrate, however, that in the context of nineteenth century British society conical refraction was linked by Hamilton and his contemporaries to many controversial positions.

By recapturing the methodological aspects of the debate over the nature of light I have shown how conical refraction represented the type of science Hamilton, Whewell, Lloyd and other mathematicians at Cambridge and Trinity College championed. It seemed to be a counterexample to the empirical approach advocated by Brewster, Brougham and Potter. In nineteenth century Britain, however, all other disciplines looked to science for examples of the most certain parts of knowledge. Coleridge, Whewell and Mill all agreed that science was the paradigm of truth, and they sought to mod economics, politics, social theory and ethics according to their own beliefs about proper science (or vice versa).

In the contexts of Section A of the BAAS, the debate between the utilitarians and the idealists and the often uncomfortable position of the Protestant Ascendancy between the Catholics of Ireland and the English, conical refraction was much more than simply a minor event in the history of the theory of double refraction. The very fact that Hamilton decided to diverge from his earlier geometrical research and to investigate Fresnel's wave theory was conditioned by these contexts, and once discovered the successful prediction was felt to have important repercussions in all of these contexts. Thus the contexts of conical refraction should not be seen simply as background, as some stage on which scientific events take place, but as a network of interactions in which science both shapes and in shaped by philosophy, society and politics.

Notes

1. I would like to thank Piers Bursill-Hall and Simon Schaffer for many enlightening discussions about Hamilton and Olivier Darrigol and Roger Hahn for commenting on drafts of this essay. This material is based upon work supported under a National Science Foundation Graduate Research Fellowship.
2. W.R. Hamilton, 'Third Supplement to an Essay on the Theory of Systems of Rays [read 1832]', *Transactions of the Royal Irish Academy*, 17 (1837): pp. 1–144, reprinted in A.W. Conway and J.L. Synge, eds., *The Mathematical Papers of Sir William Rowan Hamilton, I: Geometrical Optics* (Cambridge, 1931), pp. 164–293, for conical refraction see pp. 277–293. For Lloyd's experimental confirmation see H. Lloyd, 'On the Phenomena presented by Light in its Passage along the Axes of Biaxal Crystals', *Philosophical Magazine*, 37 (1833): pp. 112–120; H. Lloyd, 'Further Experiments on the Phenomena presented by Light in its Passage along the Axes of Biaxal Crystals', *Philosophical Magazine*, 37 (1833): pp. 207–210; and H. Lloyd, 'On Conical Refraction', *Report of the Third Meeting of the British Association for the Advancement of Science; Held at Cambridge in 1833* (London, 1834), pp. 370–373.
3. William Whewell, 'Opening Address', *Report of the Third Meeting of the British Association for the Advancement of Science*, pp. xi–xxvi, see p. xvi.
4. G.B. Airy, *Philosophical Magazine* (June 1833), p. 420 cited in Robert P. Graves, *Life of Sir William Rowan Hamilton*, (Dublin, 1882) I, p. 637.

5. J.W. Lubbock to Hamilton (Nov 30, 1835), in Graves, *Life of Sir William Rowan Hamilton*, II, p. 170.
6. Charles Babbage, *The Ninth Bridgewater Treatise: A Fragment, Second Edition* [originally published 1838], in Martin Campbell-Kelly, ed., *The Works of Charles Babbage* (New York, 1989), IX, pp. 33–34.
7. Julius Plücker, 'Discussion de la forme générale des ondes lumineuses', *Crelle's Journal*, 19 (1839): pp. 1–44, cited in Graves, *Life of Sir William Rowan Hamilton*, I, p. 637.
8. In particular George Sarton, 'The Discovery of Conical Refraction by William Rowan Hamilton and Humphrey Lloyd (1833)', *Isis*, 17 (1932): pp. 154–70, but similar sentiments are found in all the works cited in note 9.
9. See Sarton, *ibid.*; Thomas Hankins, *Sir William Rowan Hamilton* (Baltimore, 1980), ch. 6; James G. O'Hara, 'Humphrey Lloyd (1800–1881) and the Dublin Mathematical School of the Nineteenth Century', Unpublished PhD Diss-ertation (University of Manchester, 1979), ch. 3; James G. O'Hara, 'The Prediction and Discovery of Conical Refraction by William Rowan Hamilton and Humphrey Lloyd (1832–1833)', *Proceedings of the Royal Irish Academy*, 82A (1982): pp. 231–257. Hankins does provide useful background (esp. chapters 9 and 10) but, as I will argue, misses crucial aspects of the context of the prediction.
10. O'Hara, 'The Prediction and Discovery of Conical Refraction', p. 256.
11. Hankins, *Sir William Rowan Hamilton*, pp. 94–95.
12. George G. Stokes, 'Report on Double Refraction', *British Association Report*, 1862, p. 70, cited in Hankins, *Sir William Rowan Hamilton*, p. 95.
13. As early as 1831 Airy compared the status of the wave theory of light to that of the theory of gravitation, claiming 'it is certainly true'. George B. Airy, *Mathematical Tracts*, 2nd ed. (Cambridge, 1831), p. iv.
14. Sarton, 'The Discovery of Conical Refraction', p. 154.
15. See W.R. Hamilton, 'On Caustics. Part First', *Mathematical Papers*, I, pp. 345–363; Hamilton, 'Theory of Systems of Rays. Part First', *Transactions of the Royal Irish Academy*, 15 (1828): pp. 69–174, also *Mathematical Papers*, I: pp. 1–88; Hamilton, 'Theory of Systems of Rays. Part Second', also *Mathematical Papers*,: pp. 88–106; Hamilton, 'Supplement to an Essay on the Theory of Systems of Rays', *Transactions of the Royal Irish Academy*, 16 (1830): pp. 1–61, also *Mathematical Papers*, I: pp. 107–144; Hamilton, 'Second Supplement to an Essay on the Theory of Systems of Rays', *Transactions of the Royal Irish Academy*, 16 (1831): pp. 93–125, also *Mathematical Papers*, I: pp. 145–163; Hamilton, 'Third Supplement to an Essay on the Theory of Systems of Rays', *Transactions of the Royal Irish Academy*, 17 (1837): pp. 1–144, also *Mathematical Papers*, I: pp. 164–293 (except for pp. 277–293 on conical refraction).
16. Hamilton to Coleridge [not sent], October 3, 1832, in Graves, *Life of Sir William Rowan Hamilton*, I, p. 592.
17. Hamilton to John F.W. Herschel, December 18, 1832, in Graves, *Life of Sir William Rowan Hamilton*, I, p. 627.
18. See Geoffrey Cantor, 'The Reception of the Wave Theory of Light in Britain: A Case Study Illustrating the Role of Methodology in Scientific Debate', *Historical Studies in the Physical Sciences*, 6 (1975): pp. 109–132; Henry John Steffens, *The Development of Newtonian Optics in England* (New York, 1977), pp. 137–149; Cantor, 'The Theological Significance of Ethers', in Cantor and M.J.S. Hodge, eds., *Conceptions of Ether: Studies in the History of Ether Theories, 1740–1900* (Cambridge, 1981), pp. 135–155; L.L. Laudan, 'The Medium and Its Message: A Study of Some Philosophical Controversies About Ether', in Cantor and Hodge, *Conceptions of Ether*, pp. 157–185; Jack Morrell and Arnold Thackray, *Gentlemen of Science: Early Years of the British Association for the Advancement of Science* (Oxford, 1981), pp. 466–472; Cantor, *Optics After*

Newton: Theories of Light in Britain and Ireland, 1704–1840 (Manchester, 1983), esp. pp. 147–187; Xiang Chen and Peter Barker, 'Cognitive Appraisal and Power: David Brewster, Henry Brougham, and the Tactics of the Emission-Undulatory Controversy During the Early 1850s', *Studies in History and Philosophy of Science*, 23 (1992): pp. 75–101.
19. For the Scottish methodological tradition see L.L. Laudan, 'Thomas Reid and the Newtonian Turn of British Methodological Thought', in Robert E. Butts and John W. Davis, eds., *The Methodological Heritage of Newton* (Oxford, 1970), pp. 103–131; Geoffrey Cantor, 'Henry Brougham and the Scottish Methodological Tradition', *Studies in History and Philosophy of Science*, 2 (1971): pp. 69–89; E.W. Morse, '"Natural Philosophy, Hypotheses and Impiety", Sir David Brewster Confronts the Undulatory Theory of Light', Unpublished PhD Thesis (University of California – Berkeley, 1972); Richard Olson, *Scottish Philosophy and British Physics, 1750–1880: A Study in the Foundations of the Victorian Scientific Style* (Princeton, 1975).
20. John Robison, 'Philosophy', *Encyclopaedia Britannica*, 3rd ed., II, p. 593b, no. 66, cited in Morse, 'Natural Philosophy, Hypothesis and Impiety', p. 28.
21. David Brewster, 'The Revelations of Astronomy', *North British Review*, 6 (1847): p. 238, cited in Morse, 'Natural Philosophy, Hypothesis and Impiety', p. 54.
22. Whewell, 'Opening Address', p. xx (emphasis in original).
23. Whewell, *The Philosophy of the Inductive Sciences*, in G. Buchdahl and L.L. Laudan, eds., *The Historical and Philosophical Works of William Whewell* (London, 1967), V and VI. On Whewell see especially Menachem Fisch, *William Whewell, Philosopher of Science* (Oxford, 1991); Menachem Fisch and Simon Schaffer, eds., *William Whewell: A Composite Portrait* (Oxford, 1991) and Richard Yeo, *Defining Science: William Whewell, Natural Knowledge, and Public Debate in Early Victorian Britain* (Cambridge, 1993).
24. On Hamilton's philosophy of science see Graves, *Life of Sir William Rowan Hamilton*, passim; Hankins, *Sir William Rowan Hamilton*, esp. ch. 12; Fisch, *William Whewell*, pp. 63–67.
25. Hamilton, Journal entry for September 10, 1839, in Graves, *Life of Sir William Rowan Hamilton*, II, pp. 303–4 (emphasis in original).
26. David Brewster, 'Observations on the Absorption of Specific Rays, in Reference to the Undulatory Theory of Light', *Philosophical Magazine* 2 (1833): 360–61, cited in Hankins, *Sir William Rowan Hamilton*, p. 151.
27. Cited in Hamilton to Helen Bayly, March 14, 1833, in Graves, *Life of Sir William Rowan Hamilton*, II, p. 26 (emphasis in original).
28. Morrell and Thackray, *Gentlemen of Science*, p. 479.
29. Aubrey De Vere to Hamilton, December 29, 1832, in Graves, *Life of Sir William Rowan Hamilton*, II, p. 17.
30. Aubrey De Vere to Hamilton, October 6, 1832, in Graves, *Life of Sir William Rowan Hamilton*, I, p. 616.
31. On the politics of Romanticism (especially Wordsworth and Coleridge) see Crane Brinton, *The Political Ideas of the English Romanticists* (Oxford, 1926); Alfred Cobban, *Edmund Burke and the Revolt Against the Eighteenth Century: A Study of the Political and Social Thinking of Burke, Wordsworth, Coleridge and Southey* (London, 1929); Charles R. Sanders, *Coleridge and the Broad Church Movement* (Durham, North Carolina, 1942); John Colmer, *Coleridge: Critic of Society* (Oxford, 1959); R.W. Harris, *Romanticism and the Social Order* (London, 1969); Ben Knights, *The Idea of the Clerisy in the Nineteenth Century* (Cambridge, 1978); Marilyn Butler, *Romantics, Rebels and Reactionaries: English Literature and its Background 1760–1830* (Oxford, 1981); John T. Miller, *Ideology and Enlightenment: The Political and Social Thought of Samuel Taylor Coleridge* (London, 1987). On Hamilton's friendship with Wordsworth see George

Dodd, 'Wordsworth and Hamilton', *Nature*, 228 (1970): 1261–1263 as well as Hankins, *Sir William Rowan Hamilton*, *passim*; and Graves, *Life of Sir William Rowan Hamilton*, I–III, *passim*.
32. J.S. Mill, 'Coleridge' [originally published 1840], in Alan Ryan, ed., *Utilitarianism and Other Essays: J.S. Mill and Jeremy Bentham* (London, 1987), pp. 177–226, see p. 180.
33. On idealists vs. utilitarians see Sheldon Rothblatt, *The Revolution of the Dons: Cambridge and Society in Victorian England* (New York, 1968), pp. 97–116, and Yeo, *Defining Science*, pp. 176–230.
34. See Elie Halévy, *The Growth of Philosophic Radicalism*. Trans. A.D. Lindsay (London, 1928).
35. J.S. Mill, 'Bentham' [originally published 1838], in Ryan, ed., *Utilitarianism and Other Essays*, pp. 132–175, see p. 140.
36. See also James Mill, *Analysis of the Phenomena of the Human Mind*, 2nd ed. (London, 1878).
37. J.S. Mill, *Autobiography* [originally published 1873] (New York, 1957), p. 175.
38. See, for example, T.R. Birks, *On the Analogy of Mathematical and Moral Certainty* (Cambridge, 1834).
39. See S.T. Coleridge, *On the Constitution of the Church and State, According to the Idea of Each* [originally published 1829], ed. John Colmer (London, 1976).
40. In addition to Yeo, *Defining Science*, see Perry Williams, 'Passing on the Torch: Whewell's Philosophy and the Principles of English University Education', in Fisch and Schaffer, eds., *William Whewell*, pp. 117–147.
41. J.S. Mill, 'Whewell on Moral Philosophy [originally published 1852]', pp. 228–270 in Ryan, ed., *Utilitarianism and Other Essays*, p. 230.
42. Eliza Mary Hamilton, 'Wordsworth at the Observatory, Dunsink', (Aug 1829), in Graves, *Life of Sir William Rowan Hamilton*, I, p. 313.
43. See J.N. Hays, 'Science and Brougham's Society', *Annals of Science*, 20 (1964): 227–241; Harold Smith, *The Society for the Diffusion of Useful Knowledge, 1826–1846: A Social and Bibliographic Evaluation* (Halifax, Nova Scotia, 1974) and Colin A. Russell, *Science and Social Change in Britain and Europe, 1700–1900* (New York, 1983), pp. 136–73.
44. Coleridge, *On the Constitution of the Church and State*, p. 69.
45. J.S. Mill, *Autobiography*, p. 145.
46. I do not want to argue that Hamilton simply 'made up' conical refraction. What I am interested in is why Hamilton would look for such a phenomenon and what the phenomenon once discovered was taken to mean or imply.
47. On mathematics at Cambridge and Trinity College in the early nineteenth century see, for example, John Purser, 'President's Address to the Mathematical and Physical Science Section', *Report of the Seventy-Second Meeting of the British Association for the Advancement of Science; Held at Belfast in 1902* (London, 1903), pp. 499–511; A.J. McConnell, 'The Dublin Mathematical School in the First Half of the Nineteenth Century', *Proceedings of the Royal Irish Academy*, 50 (1944): 75–88; Maurice Crosland and Crosbie Smith, 'The Transmission of Physics from France to Britain: 1800–1840', *Historical Studies in the Physical Sciences*, 9 (1978): 1–61; N.D. McMillan, 'The Analytical Reform of Irish Mathematics 1800–1831', *Newsletter of the Irish Mathematical Society*, 10 (1984): 61–75; I. Grattan-Guinness, 'Mathematics and Mathematical Physics from Cambridge, 1815–40: A Survey of the Achievements and of the French Influences', in P.M. Harman, ed., *Wranglers and Physicists: Studies on Cambridge Physics in the Nineteenth Century* (Manchester, 1985), pp. 84–111; I. Grattan-Guinness, 'Mathematical Research and Instruction in Ireland, 1782–1840', in John Nudds, Norman McMillan, Denis Weaire, and Susan McKenna Lawlor, eds., *Science in Ireland 1800–1930: Tradition and Reform. Proceedings of an International*

Symposium held at Trinity College Dublin March 1986 (Dublin, 1988), pp. 11–30; N. Guicciardini, *The Development of the Newtonian Calculus in Britain 1700–1800* (Cambridge, 1989), pp. 95–142.
48. Hamilton to Aubrey De Vere, February 9, 1831, in Graves, *Life of Sir William Rowan Hamilton*, I, p. 519.
49. On Hamilton's philosophy of mathematics see esp. Thomas L. Hankins, 'Algebra as Pure Time: William Rowan Hamilton and the Foundation of Algebra', in P.K. Machamer and R.G. Turnbull, eds., *Motion and Time, Space and Matter* (Ohio, 1976), pp. 327–359; David Bloor, 'Hamilton and Peacock on the Essence of Algebra', in H. Mehrtens, H. Bos and I. Schneider, eds., *Social History of Nineteenth-Century Mathematics* (Boston, 1991), pp. 202–232; Anthony T. Winterbourne, 'Algebra and Pure Time: Hamilton's Affinity with Kant', *Historia Mathematica*, 9 (1982): 195–200; John Hendry, 'The Evolution of William Rowan Hamilton's View of Algebra as the Science of Pure Time', *Studies in the History and Philosophy of Science*, 15 (1984): 63–81; Peter Ohrstrøm, 'W.R. Hamilton's View of Algebra as the Science of Pure Time and His Revision of this View', *Historia Mathematica*, 12 (1985): 45–55. For early nineteenth century British debates about the foundations of algebra see also Joan L. Richards, 'The Art and the Science of British Algebra: A Study in the Perception of Mathematical Truth', *Historia Mathematica*, 7 (1980): 343–365; Helena M. Pycior, 'Early Criticism of the Symbolical Approach to Algebra', *Historia Mathematica*, 9 (1982): 392–412; Helena M. Pycior, 'Internalism, Externalism, and Beyond: 19th-Century British Algebra', *Historia Mathematica*, 11 (1984): 424–441; Menachem Fisch, '"The Emergency Which Has Arrived": The Problematic History of Nineteenth-Century British Algebra – A Programmatic Outline', *British Journal for the History of Science*, 27 (1994): 247–76.
50. See J.C. Beckett, *The Anglo-Irish Tradition* (London, 1976) and W.J. McCormack, *Ascendancy and Tradition in Anglo-Irish Literary History from 1789–1939* (Oxford, 1985).
51. See R.B. McDowell and D.A. Webb, *Trinity College Dublin 1592–1952: An Academic History* (Cambridge, 1982).
52. R.F. Foster, *Modern Ireland, 1600–1972* (London, 1989), p. 173.
53. See R.B. McDowell, *Ireland in the Age of Imperialism and Revolution, 1760–1801* (Oxford, 1979).
54. See R.B. McDowell, 'The Main Narrative', in T. O'Raifeartaigh, ed., *The Royal Irish Academy: A Bicentennial History, 1785–1985* (Dublin, 1985), pp. 1–92.
55. See Patrick A. Wayman, *Dunsink Observatory, 1785–1985: A Bicentennial History* (Dublin, 1987).
56. See James Bennett, *Church, State and Astronomy in Ireland: 200 Years of Armagh Observatory* (Armagh, 1990).
57. See Oliver MacDonagh, *Ireland: The Union and Its Aftermath* (London, 1977).
58. Hamilton to H. Lloyd, January 16, 1836, in Graves, *Life of Sir William Rowan Hamilton*, II, p. 177.
59. Aubrey De Vere to Hamilton, December 29, 1832, *ibid.*, II, p. 17.
60. Hamilton to Aubrey De Vere, February 6, 1847, *ibid.*, 2, p. 558.
61. See, for example, Hamilton speech to the 1832 meeting in Oxford, *ibid.*, 1, p. 571.
62. See Fergus O'Ferrall, *Catholic Emancipation: Daniel O'Connell and the Birth of Irish Democracy 1820–30* (Dublin, 1985) and Kevin B. Nowlan, *The Politics of Repeal: A Study in the Relation Between Great Britain and Ireland, 1841–50* (London, 1965).
63. *The Dublin Evening Mail* No. 1799, Wed, Aug 20, 1834.
64. Sarton, 'The Discovery of Conical Refraction', p. 156.

Chapter 3

Science and Social Policy in Ireland in the Mid-Nineteenth Century

James Bennett

Science in nineteenth-century Ireland has become an increasingly attractive area of study for historians. The former focus on science's relation, in Britain and elsewhere, to industrial development, has for some time now made room for studies of its role in forging social and cultural identities. In this context, Irish case studies seem promising because of the significance of science for cultural differentiation and, by the same token, its potential as a tool for social and educational policy. This paper offers first a grounding for our appreciation of the cultural identity of science in Ireland, through a review of the community of scientists, and then an example of how the issues that divided them were played out in an important debate in the mid-century, namely the establishment of alternative institutions of higher education to Trinity College and the foundation of the Queen's Colleges.

The basis for the paper is a survey of aspects of the biographies of Irish scientists and scientists working in Ireland in the nineteenth century. If we search the first four volumes of Poggendorff's *Biographisch-Literarisches Handwörterbuch zur Geschichter der Exacten Wissenschaften* for scientists with strong Irish connections – birth and education or part education, or career – a sample of around 200 emerges. For readers not familiar with Poggendorff, inclusion indicates a significant standing based on publications. The sample is sufficient to support a social profile of science in nineteenth-century Ireland, and much of what emerges is as might have been expected.[1]

We can pass over confirmations of expectations fairly quickly. For example, Britain is confirmed as the main external cultural influence on science in Ireland, at least as mediated by the careers of her scientists. In the sample, 25% were born in Britain, 8% on the Continent and 3% elsewhere. Of those born in Ireland, or of Irish parents temporarily abroad, 11% were wholly and a further 25% partly educated in Britain. By contrast with this total of 36%, only 12% were wholly or partly educated on the Continent. We might note that most of them were chemists: chemistry seems to have been a stronger factor in this respect than Catholicism.

Of those who had already sufficient connection with Ireland to be sampled before embarking on a career, 22% subsequently spent their entire careers in Britain or in the Empire, while 36% of the entire sample divided their careers between Britain and Ireland. By contrast, only 5% pursued careers that included positions on the Continent and 4% in America.

In connection with this group of results it is worth pointing out that 35% of the entire sample pursued their careers wholly in Ireland, while 22% fall into the category of those born in Ireland, wholly educated in Ireland and having their entire careers in Ireland. Given the geographical and historical circumstances, these are quite high figures and point to a strong scientific community.

Another unsurprising group of results concerns religious persuasion. Although there was a significant Catholic presence, science was very largely the cultural domain of the Protestant Ascendancy: it was middle-class and professional and in the southern provinces Anglican, but there was a strong nonconforming representation in Ulster.

We can attach some figures to those results. Of the members of the sample who were born in Ireland, I have been able to establish the faiths of 87% and of these 62% were Church of Ireland, 10% Nonconformists, 17% Protestants about whom I could not be more precise and 10% were Roman Catholics. It could be said that the Catholic representation played a more significant role than this might indicate because a number of its members – such as John Casey, Henry Hennessy, William Higgins, Robert Kane and William K. Sullivan – were important figures. There are also significant examples, such as Nicholas Callan and Gerald Molloy, who do not appear in Poggendorff.[2]

The social origin of the Irish-born members of the sample shows very little differentiation. They were almost all from the middle ranks of society – the sons of merchants, industrialists, professionals and minor landed gentry. A noticeable number were the sons of Church of Ireland clergymen, or had such clergymen for maternal grandfathers. There are very few representatives of the aristocracy, almost as few of the working class and no women.

If we try to gauge the politics of members of the sample, any pretence at quantification must be dismissed, but some general remarks are still possible. Views on such national issues as the Union, emancipation, repeal, the universities, disestablishment, home rule, and so on are widely voiced, and a good many scientists had a significant involvement in public life. Examples would be Thomas Drummond, Richard Griffith, John Pope Hennessy, Robert Kane, William K. Sullivan, John Ball, Charles J.

Hargreave, William James MacNeven, Alexander Nimmo, George Salmon and George Johnstone Stoney.[3] From the end of the eighteenth century to the third quarter of the nineteenth there is a striking range of positions on relevant issues. There are United Irishmen and sympathisers, such as MacNeven, Richard Kirwan and James Thomson, reformers such as Drummond and nationalists like James McCullagh and John Birmingham.[4] There are also such staunch conservatives as William Hales, Edward J. Cooper, Wentworth Erck and Thomas Romney Robinson.[5]

Political diversity is sustained until towards the end of the century, but among Irish scientists in middle life around 1890, there are moderate views to be found but nothing really radical. Rather, there is a list of convinced Unionists, such as R.S. Ball, Joseph Larmor, John Joly and the fourth Earl of Rosse.[6] There is no particular significance to this growth of unionism: it reflects trends in the Protestant population as a whole, to which most leading scientists still belonged. Overall there is certainly no evidence of congruence between political views and the practice of science in Ireland.

After these anticipated and unexceptional results, there is one particularly striking one, namely the preponderance of professionals in Irish science. The picture which emerged was quite different from the 'Gentlemen of Science' characterisation which Morrell and Thackray have given to England at a similar time. According to them the influential British Association for the Advancement of Science was hostile to equal participation 'by women, workers, provincials, or professionals not committed to gentlemanly voluntarism'.[7] In Ireland 86% of the scientists mentioned in Poggendorff for the nineteenth century were professional in the sense that they were in paid employment directly related to science. Some of the 14% of amateurs were university professors holding chairs in non-scientific subjects.

The very high incidence of professional careers reflects the degree to which the scientific enterprise was controlled by public institutions and through them by a central administration. Through a whole series of parliamentary commissions and more permanent government agencies, notably the Ordnance Survey and the Geological Survey, a scientific career could be sustained in the public service in a variety of ways. Government exercised greater control over higher education in Ireland than in England, the Queen's Colleges being founded with deliberate social and political intentions. Other institutions, such as the Museum of Irish Industry, the Department of Science and Arts and the Royal College of Science contribute to a picture of significant central influence. Irish scientific societies were receiving annual grants from the government considerably before their counterparts in England. The Dublin Society began receiving generous grants from the Irish Parliament as early as 1761, and by the beginning of the Union an annual grant of £10,000 was being paid.[8] This exposed the society to government manipulation, through a whole series of select committees and commissions. Recommendations were often enforced by the threat of withholding a grant, so that the history of the society in the period is a catalogue of centrally-inspired changes and reforms. The Royal Irish Academy also received grants which, though considerably smaller than those of the society, had become annual by 1800, through one of the last legislative measures of the Irish Parliament. After 1816 this grant became £300 per annum, it was raised to £500 in 1855 and to £1500 in 1864. All the while the grants entailed regular government scrutiny and periodic administrative upheaval, or the threat of upheaval.[9]

Thus distinctive features of science in nineteenth-century Ireland are its association with a particular cultural identity or social interest, an association which was obvious to those concerned, and the manipulation of its institutions by a power whose aims did not necessarily coincide with those of any of the parties directly involved. While it might seem dubious to translate the analysis of an artificial sample into sentiments in play at the time, there is sufficient evidence that these conclusions coincide with the understanding of Irish scientists. In response to a charge that Catholics, were not successfully involved with science, the chemist William K. Sullivan was stung to retort:

> Is it not a mockery for a member of that ascendancy party, which used in former times such unholy means to crush out every trace of mental culture from amongst us, and who now uses mean calumny and vulgar gibes, to ask us where are our senior wranglers?[10]

Are there models from other examples where science is part of a pattern of cultural distinction, which might be helpful to the historian thinking about its operation in Ireland? The Irish situation is complicated by there being several cultural constituencies, whose allegiances and connecting interests are not fixed.

One model from elsewhere that cannot apply to Ireland is that of science cultivated by a minority excluded from power, influence and mainstream education, such as has been used in connection with seventeenth-century English puritans or eighteenth-century English dissenters. More promise seems to he offered by the model of science used as the cultural property of a powerful section of society to enhance its claims to the guardianship of enlightened cultural interests and to promote its particular ethos. Examples where this has been used would be the relation between Newtonian natural philosophy and latitudinarian churchmen in England around 1700 or the ethos of Anglican liberalism characteristic of the early British Association's 'Gentlemen of Science'.

While the latter model seems more promising, it cannot give a complete account, for the practice of science in Ireland was not so firmly based on a voluntary code as it was in England. It was to a large extent promoted and manipulated by a power outside the country, and manipulated to a degree that would not have been possible in England. The kinds of voluntary associations whose common interests might be sustained and forwarded through science certainly existed in nineteenth-century Ireland, but there existed also an external force in the London administration with its representation in Dublin, which used science and its promotion as an instrument for change and social control, and this was felt directly by those very organisations.

This external force had its own model, a way of characterising science which it sought to promote. This represented science as a uniting influence in a divided society, encouraging men to forget their differences in its pursuit, to join in an enterprise that transcended factions and to learn to live at peace. This was not necessarily the political motivation of the Peel administration as it sought to accommodate the Catholic majority in Ireland, but it was certainly employed in official rhetoric. Nor was its use original or novel. In England it goes back at least to Thomas Sprat's apology for the early Royal Society, where he explained that they sought 'not to lay the Foundation of an English, Scotch, Irish, Popish, or Protestant Philosophy; but a Philosophy of Mankind'.[11]

A single example will bring out some of the variety and complexity of the interests at work. This concerns the foundation of the Queen's Colleges in the mid-century, the debate between mixed and exclusive higher education, and the role of science.

Was science a cultural property? Certainly in Ireland the realistic answer had to be 'yes'. Was this an inevitable state of affairs, so that if another section of Irish society was to be admitted to the practice of science and was to retain its identity, this science would have to be sponsored by a different culture, and would this make for a different practice? Or, despite its record in Ireland, could science somehow transcend faction, in which case the appearance of exclusivity would have to be understood as a social artefact of no importance to the nature of science. All of these positions were in play in the university education debate, which makes it of particular interest to current issues in the history of science.

The government's response to the need for an alternative to the Anglicanism of Trinity College was not to found and endow a Catholic university for Catholics and a Presbyterian university for Presbyterians, but to establish the Queen's Colleges in Cork, Galway and Belfast, without any denominational basis and without state-endowed theological chairs.[12] In some respects this was an extension of the National School system, which was founded on a mixed or universal basis, even though most of the schools became in practice denominational.

With the establishment of the Queen's Colleges there began a protracted debate, in which the leading scientists in public life were frequently involved. It was a debate over the morality of the colleges, their effectiveness and success, in what form they might continue and how they might be replaced, while all the time the demand for a truly Catholic institution was maintained. The scientists who contributed notably included Robert Kane, who became President of Queen's College Cork and who was founder and Director of the Museum of Irish Industry, William K. Sullivan, Kane's assistant at the Museum, Henry Hennessy, Professor of Physics at the Catholic University after being Librarian at Queen's College, Cork, and Samuel Haughton, Professor of Geology at Trinity College.

Science was centrally involved in this debate, for it was an important part of the teaching programme and it was around science that most of the charges of godlessness and secularity were focused. The widespread, though far from unanimous, opposition to the colleges among the Catholic clergy is fairly well known, but concern about the role of science in the supposed spread of irreligion was not confined to Catholic sensibilities. Thomas Romney Robinson regaled the annual meetings of the Armagh History and Philosophical Society about 'the stream of unbelief, which is now running so freely', and he used contemporary geology and evolutionary theory to support this fear. Such resistance did not denote opposition to science or a lack of understanding or of sympathy with the scientific enterprise: Robinson was one of the most prominent public figures in the science of his day, in Britain as well as in Ireland.[13] At issue rather was the proper context for science education: if science was dangerous outside the enclosing context of religion, the Queen's Colleges were the vehicle by which science might precipitate a collapse of religious values.

Science was a very significant part of the curriculum at the colleges from the beginning: natural philosophy, chemistry, mineralogy, geology and natural history were all prescribed, and provision was made for laboratories and museums. It is notable, however, that students came to natural philosophy only in their third year, after they had challenged themselves on the rigours and certainties of classics and mathematics. Chemistry and natural history came in the second year, while the full course of physical science was available only at the master's level. This was a sensibility addressed also, for example, at Cambridge, where college tutors were concerned about the effect on the young of the insecure and contingent regime of experimental physics. The early annual reports from the Queen's Colleges laid great stress on the morals and religious observance of the students, as well as emphasising the harmony between students of different denominations.[14] The numbers of students of each persuasion were given, to demonstrate the application of mixed education.

Among the scientists who contributed in a significant way to the argument, Haughton was the only real Trinity man, and in general Trinity maintained a distance from the debate over the Queen's Colleges. While it was an important component in the situation which gave rise to the need for alternative institutions, the decision to found the colleges signalled that, for the present at least, Trinity had been removed from the question. Haughton's views were moderate and they were published in 1868, rather later than the main thrust of debate. He wanted the character of Trinity to remain unchanged and advocated the foundation and endowment of a Catholic University.[15]

The argument against a Catholic University and the political justification for promoting mixed rather than denominational institutions, rested on the assertion that it was in sectarianism that the roots of the Irish problems lay and that non-sectarian education was an enlightened and liberal attempt to rise above petty differences through the common cultivation of science and letters. This self-righteous liberalism was neatly turned around by the political opposition, who said that true liberality would allow institutions catering for every persuasion. It was further asserted that the new high-minded liberalism was a hypocritical gloss on a subtle form of social control. As William K. Sullivan wrote in 1866:

> ... the ascendancy party in Ireland ... knows that a properly educated middle class would soon deprive them of a monopoly which they formerly defended in the name of conservatism, but which they now propose to maintain in the name of liberalism and enlightenment.
>
> In former times Irish Protestants memorialled to deprive us of all education. But things have so progressed in the 'nineteenth century', they only demand that we be preserved from educating ourselves; they regret the thinness of our intellectual food, and so would give us a substantial mental pudding of their own making.[16]

Henry Hennessy had made the point some years earlier. In 1858 he had explained to an audience in England that

> ... in Ireland ... at this moment, pretensions to the exclusive possession of liberal ideas, are most loudly made by those who, at the same time, insist on forcing the whole system of public instruction into strict conformity with their own peculiar views.[17]

The speech was published at a time when the colleges were under particular attack. In the same year, Henry Hennessy's brother, John Pope Hennessy, published a tract entitled 'The Failure of the Queen's Colleges' – a failure, he said, 'total and complete'.[18] His critique included the very serious assertion that the educational structure of the colleges allowed revealed religion to be undermined through the teaching of science outside any denominational framework.

Three general positions on the issue are discernible among scientists in the mid-nineteenth century. A pragmatic, liberal stance was adopted by reforming Catholics, broadly in favour of the establishment of the colleges and their regimes of education. Their principal spokesman was Robert Kane, whose programme was one of extending higher education to the middle classes of all persuasions through mixed institutions, and of social and material progress based on the recruitment of applied science and the development of industry in Ireland.

The lobby for a Catholic University argued either that the Queen's Colleges were godless, or that they were part of a political and social conspiracy of the Establishment, designed to undermine the Catholic tradition. Henry Hennessy might be the best representative of this position.

The Established Protestant position was to claim a transcendental role for science, above the petty differences and grubby political realities of Ireland. Such sentiments might be presented in an extraordinary rhetoric. For example, George Boole, an Englishman who found himself uncomfortably involved in this Irish controversy by virtue of his appointment to the mathematics chair at Cork, was convinced of the justice of the official position and that it would eventually triumph. It was a confidence he expressed with almost religious sentiment in a public lecture in the college in 1851, where he said, with specific reference to the situation in Ireland:

> Truth shall assert her rightful claims. Science shall vindicate her divine mission in the increase of the sum of human good. Obscured by the mists of prejudice, forgotten amid the strife of parties, she but the more resembles those great luminaries of heaven, which pursue their course undismayed above the rage of tempests, or amid the darkness of eclipse.[19]

Another example of this rhetoric would be Lord Rosse's presidential address to the British Association at Cork in 1843, where, again with specific reference to the situation in Ireland, he says:

> Each successive discovery, as it brings us nearer to first principles, opens out to our view a new and more splendid prospect, and the mind, led away by its charms, is carried beyond and far above the petty and ephemeral contests of this life . . .
>
> . . . the religion of discovery is rich beyond the powers of contemplation; and however much we may draw from it we shall not leave its treasures exhausted – no, not even diminished, because they are infinite.[20]

By setting 'the religion of discovery' in place of 'petty and ephemeral contests', the Establishment sought to place the enterprise of science on a higher moral ground, from where present problems were seen as irrelevant to its search for truth. It is interesting that both these representatives of the vision invest it with aspects of the divine.

There are many ways in which the history of science in Ireland is significant for current approaches to the subject, but one is surely indicated by this controversy. At its heart is the question of whether science can claim a disinterested agenda, or whether it is always socially constructed, so that such an appeal is itself socially and politically interested, as its opponents in Ireland claimed it was. The curious political and social role of science in Ireland and the recognition of this by everyone concerned, enforced the question of whether science could transcend cultural interest.

As a representative of those who wished to use the potential of science as social policy to further their political agendas, we can take Kane as the most eloquent. In his inaugural address as President, at the opening of Queen's College, Cork, in 1849, he put the case for what he called 'the principle of free, and impartial, and united education.'

> ... these colleges are founded for this country and for its people; not for a party nor for a class, not for an ascendancy not for a creed; but that, in the pure and soul-ennobling paths of intellectual glory, all ranks, all sects, all parties of the Irish people may unite – may learn to know and love each other ... and learn to act in harmony and concert ... From age to age we have been forced to see different elements of our population reared up in mutual ignorance, separated by barriers of social instincts – strengthened by misdirected education. Let us have done with this. Let us, at least in the calm retreats of literature and science ... render available to our general people, those privileges of study from which we have been so long debarred.[21]

For an expression of the contrary view, with a similar rhetorical quality, we can turn to Henry Hennessy, speaking at the inauguration of the faculty of science in the privately-established Catholic University. The speech was a difficult assignment, as Hennessy had to explain that science, despite its obvious associations in Ireland, was an appropriate part of a truly Catholic education, that it should not be regarded as un-Catholic on either religious or cultural grounds.

> ... is it creditable to ourselves, or to our country, that these pursuits [the sciences] should be systematically cultivated, almost exclusively, by one portion of the population? ... we have not as yet attempted to assert our unquestionable right to cultivate science in halls of our own, under the protection and with the sanction of our religion. Who is there that will venture to deny our right to cultivate knowledge in our own way? ...

> ... If our claim was not made before, the delay is easily explained by a glance over the history of our country. Let us turn from these recollections – let us look forward to the future that must dawn on our island, when its entire population shall be conscious of educational equality, when no direct or indirect disability shall obstruct those who wish to pursue the higher branches of knowledge in establishments guided, sanctioned, and vivified by religion.[22]

The present article can do no more than indicate that this fascinating debate goes to the heart of current concerns in the history and philosophy of science, and suggest that Sullivan would be one figure worth a much closer study, on account of his intimate involvement with the debate and with the relevant institutions, and of the shifts in his career.[23]

Sullivan's early career was spent under Kane, as Professor of Chemistry at the Museum of Irish Industry. He seems to have subscribed to Kane's programme at this stage. For example, his editorship of the failed monthly *Journal of Progress*, with its twin sections, *Journal of Industrial Progress* and *Journal of Social Progress*, indicates his assent to Kane's agenda of industrial development in Ireland linked to educational reform. However, he was to become Professor of Chemistry at the Catholic University, an appointment which antagonised his superior at the Museum, and still later he became Kane's successor at Cork.

Many of the issues surrounding the social function of science in Ireland and its cultivation in particular cultural and religious contexts are present in Sullivan's writings. He confirms the Catholic sense of exclusion from science:

With the whole field of the physical sciences, the great element of modern philosophic discussion, and intellectual progress, closed by law or prejudice against Catholics – indeed with the whole field of all science closed against them, it was unlikely that many would devote themselves to their cultivation.[24]

He analyses the distribution of appointments among scientists of different denominations, and comes to similar conclusions to those drawn in this article from a larger sample. At this point in his career, in 1866, he judges that the Queen's Colleges have failed to solve the problem, that they have not provided an environment for the higher education of Catholics. While he looks now to the Catholic University for such an environment, he wonders what this can mean for the nature of science:

> It is quite true, that there is no Catholic geology or mathematics, nor Protestant chemistry; the physical sciences have their own methods of induction from well established facts, which the experience of any one can incontrovertibly test, and from these methods there can be no deviation without error.[25]

So, he is not a social constructivist. Yet, he believes that science should be taught to Catholic students in institutions with a Catholic ethos and staff, so as to eschew secularity and godlessness, and to give science a new cultural identity in Ireland.

The debate over the appropriate institutions of higher education is an example of how the curious situation of science in nineteenth-century Ireland presents us with a promising case study relevant to much wider concerns in the history of science. One of the particularly interesting aspects is the extent to which the scientists themselves were explicit about the operation of science within a complex but relatively well-defined cultural map. It seems likely that other episodes in the nineteenth century, perhaps the period when an energetic and assertive community of scientists in Ireland had its highest profile in the European context, will yield further studies with similar implications.

NOTES

1. In addition to the standard works of reference, such as the *Dictionary of National Biography* and the obituaries published by scientific societies, of particular use have been, R.B. McDowell and D.A. Webb, *Trinity College Dublin 1592–1952: an Academic History* (Cambridge, 1982) and G.L. Herries Davies, *The History of Irish science: a Select Bibliography* (Dublin, 1985). Note also G.L. Herries Davies, *Sheets of Many Colours. The Mapping of Ireland's Rocks* (Dublin, 1983); C. Mollan, W. Davis and B. Finucane, *Some People and Places in Irish Science and Technology* (Dublin, 1985) and *More People and Places in Irish Science and Technology* (Dublin, 1990); J.H. Andrews, *A Paper Landscape: the Ordnance Survey in Nineteenth-Century Ireland* (Oxford, 1975); J.E. Burnett and A.D. Morrison-Low, *'Vulgar and Mechanick': the Scientific Instrument Trade in Ireland 1650–1921* (Edinburgh and Dublin, 1989); J. Nudds *et al.*, eds., *Science in Ireland 1800–1930: Tradition and Reform* (Dublin, 1988).

2. P.J. McLaughlin, *Nicholas Callan, Priest-Scientist, 1799–1864* (Dublin and London, 1965); C. Mollan and J. Upton, *St Patrick's College, Maynooth. The Scientific Apparatus of Nicholas Callan and other Historic Instruments* (Maynooth and Dublin, 1994).
3. Apart from standard biographical sources, note J. Ball, *What is to be done for Ireland?* (London, 1849).
4. J. Birmingham, *Anglicania, or England's Mission to the Celt* (London, 1863); P.J. McLaughlin, 'Richard Kirwan: 1733–1812', *Studies*, 28 (1930): pp. 461–474, 593–605; 29 (1940): pp. 71–83, 281–300; W.J. MacNeven, *Memoire, a Detailed Statement of the Origin and Progress of the Irish Union* (London, 1802); 'Death of Professor McCullagh', *The Nation*, 30 October 1847, p. 889.
5. W. Hales, *Observations on Tithes* (London, 1794); W. Hales, *Methodism Inspected* (Dublin and London, 1803); W. Erck, *The Land Question* (Dublin, 1883); for Robinson, see J.A. Bennett, *Church, State and Astronomy in Ireland: 200 Years of Armagh Observatory* (Armagh and Belfast, 1990), ch. 8.
6. W.V. Ball, ed., *Reminiscences and Letters of Sir Robert Ball* (London, 1915); P.A. Wayman, *Dunsink Observatory, 1785–1985: a Bicentennial History* (Dublin, 1987); J. Joly, *Reminiscences and Anticipations* (London, 1920).
7. J. Morrell and A. Thackray, *Gentlemen of Science. Early Years of the British Association for the Advancement of Science* (Oxford, 1983), p. 28.
8. H.F. Berry, *A History of the Royal Dublin Society* (London, 1915), pp. 209–214.
9. T. Ó Raifeartaigh, ed., *The Royal Irish Academy: a Bicentennial History* (Dublin, 1985), pp. 21–56.
10. W.K. Sullivan, *University Education in Ireland. A Letter to Sir John Dalberg Acton* (Dublin, 1866), p. 15.
11. T. Sprat, *The History of the Royal Society of London*, 2nd ed. (London, 1702), p. 63.
12. For Queen's College Belfast, se T.W. Moody and J.C. Beckett, *Queen's Belfast 1845–1949: The History of a University* (London, 1959). Since the conference in Armagh, there has appeared J.A. Murphy, *The College: A History of Queen's/University College Cork, 1845–1995* (Cork, 1995).
13. Bennett, *Church, State and Astronomy in Ireland*, chs 4–8.
14. These can be found in Parliamentary Reports from 1850 on; they are indexed in P. Cockton, *Subject Catalogue of the House of Commons Parliamentary Papers 1801–1900* (Cambridge, 1988), vol. 4, pp. 535–539.
15. S. Haughton, *University Education in Ireland* (London, 1868). Note also H. Lloyd, *The University of Dublin in its Relations to the Several Religious Communities* (Dublin, 1868).
16. Sullivan, *University Education in Ireland*, pp. 5–6.
17. H. Hennessy, *On Freedom of Education* (Dublin, 1859), p. vi.
18. J.P. Hennessy, *The Failure of the Queen's Colleges, and of Mixed Education in Ireland* (London, 1859).
19. G. Boole, *The Claims of Science, Especially as Founded in its Relation to Human Nature* (London, 1851), p. 30.
20. C. Parsons, ed., *The Scientific Papers of William Parsons, third Earl of Rosse 1800–1867* (London, 1926), p. 49.
21. R.J. Kane, *Inaugural Address delivered at the Opening of Queen's College, Cork* (Dublin, 1849), p. 15.
22. H. Hennessy, *On the Study of Science in its Relations to Individuals and to Society* (Dublin, 1858), pp. 47–52.
23. For an account of Sullivan, see T.S. Wheeler, 'Life and work of William K. Sullivan', *Studies*, 34 (1945): pp. 21–36. Note also T.S. Wheeler, 'Sir Robert Kane: Life and Work', *Studies*, 33 (1944): pp. 158–68, 316–30.
24. Sullivan, *University Education in Ireland*, p. 16.
25. *Ibid.*, p. 22.

Chapter 4

Science and Nationality in Edwardian Ireland

Nicholas Whyte

Part of the task of the historian of science is to find a framework which will bring a greater understanding of what happened in the past; to produce something better and more useful than a Whig positivist story about the progress of science through the careers of a series of Great Men (and it usually is men). To learn more about the science of the past we must explore the connections between scientists and their social and cultural surroundings, and the impact that these had on the kind of science that was practised in a given place at a given time.

The 'history of Irish science' brings its own potential pitfalls. If we attempt to analyse it as a unified scientific enterprise across the whole island it is unlikely that we will come up with anything meaningful, but what are the alternatives? There is a perception shared by some writers that science somehow 'failed' in post-independence Ireland because Irish nationalism is essentially Romantic, anti-materialist, anti-rationalist and therefore fundamentally anti-scientific. But my own work in a slightly earlier period has convinced me that the links between the national revival and scientific activity are sufficiently numerous for this to be too simplistic an analysis.

It is certainly true that the work of Irish scientists has not been sufficiently valued by 'mainstream' Irish historians, who have tended to concentrate on the history of Irish politics – a compliment which most of the politicians have returned by concentrating on their own versions of Irish history. John Wilson Foster has looked at a number of recent histories of Ireland and Irish studies, including Terence Brown's *Ireland: A Social and Cultural History*, Roy Foster's *Modern Ireland 1600–1972*, F.S.L. Lyons' *Ireland since the Famine*, and Joe Lee's *Ireland 1912–1985*, all of which are now standard works, and has drawn attention to the lack of any discussion of science in them.[1]

Now that other strands of Irish history – women's history, social and economic history, the history of the labour movement – are starting to emerge blinking into the light, the goal of this conference must be for the history of science in Ireland to undergo the same transition. Foster's article was published in 1991, at almost exactly the same time as a pamphlet by Margaret Ward which similarly looked at the absence of Irishwomen and gender issues from those same books.[2] She rightly bemoans the limited career prospects for women scholars who try and advance these issues from within the historical profession, but I suspect that they may yet find it easier to get jobs in their chosen discipline than do we historians of science.

In this paper, I propose a three-strand framework for the understanding of the interactions between Irish science and the other intellectual and social currents in Ireland in the years from 1900 to 1920, and examine some of the evidence supporting it. I will then look at a particular case-study of the interaction between the forces of Irish nationalism and one of the three strands within Irish science.

THREE STRANDS OF IRISH SCIENCE

My proposed framework is very loosely based on that outlined by George Basalla in his famous 1967 article in *Science*.[3] Basalla proposed a three-phase diffusionist model for the spread of Western science outside Europe. During phase one, the non-scientific society or nation provides a source for European science. The word non-scientific refers to the absence of modern Western science and not to a lack of ancient, indigenous scientific thought of the sort to be found in China or India. This means exploration by travellers, diplomats, anthropologists, botanists, zoologists, from an outside, scientifically developed culture.

Phase two is marked by a period of 'colonial science'; a small 'native' scientific community begins to appear but its work needs to be validated by scientific societies and scientists in the metropolis. Phase three completes the process with an independent scientific tradition or culture, ending up with a nation that produces and trains its own scientists who look to their own national scientific community for support and career progress.

Two hefty volumes have appeared in recent years largely filled with essays refuting the applicability of Basalla's model to particular situations.[4] It is too one-dimensional, pays too little attention to economic and social factors, is based too closely on the experience of the United States. The consensus seems to be 'It's not that simple!' or more accurately 'It's not that simple in *my* country!' One of the best known alternatives is Roy MacLeod's concept of a 'moving metropolis' rather than one which is geographically fixed; from the Irish point of view, the metropolis might sometimes be in the National History Museum in South Kensington; or the Pasteur Institute in Paris; or even in Dublin or Belfast.[5]

Although MacLeod uses the idea of the 'moving metropolis' to put forward a five-stage taxonomy of the history of science in Australia, I shall follow Basalla to the extent of limiting myself to three strands in my description of Irish science in the Edwardian era.[6] First and oldest is the tradition of science practised by the Irish Ascendancy. Although the Ascendancy are often described as 'Anglo-Irish' there was very little 'Anglo' about them; J.C. Beckett in *The Anglo-Irish Tradition* (1976) admits that he uses the term with reluctance, and both he and Roy Foster in their respective surveys of Irish history tend to use 'Ascendancy' instead.[7] Ascendancy science was validated not by the Royal Society but by the Royal Irish Academy and the Royal Dublin Society. When Irish scientists participated in the British Association it was on the same terms as their English and Scots counterparts, and certainly in its early years a fair proportion of the British Association's annual meetings were held in Ireland.

The Irish scientific tradition, especially in the two centuries before Partition, is in fact largely an Ascendancy phenomenon. This is the tradition of Trinity College, of the Royal Dublin Society, the Royal Irish Academy, and of the great telescopes. In the nineteenth century, William Rowan Hamilton, the Earls of Rosse, John Tyndall, George Francis Fitzgerald, and in this century John Joly are the most celebrated Irish scientists, and all belonged to the Ascendancy tradition. These were scientists who were established in an intellectual community of their own, whose metropolis was to be found between Trinity College, Leinster House and Dawson Street.

Ascendancy science has been comparatively well chronicled, largely by its own members and their heirs; but this has obscured the fact that it was an exclusive and to a certain extent an excluding tradition. Almost all Ascendancy scientists were Protestants (the most obvious exception being Robert Kane), and many were very hostile to Catholicism. It is particularly revealing that, although Protestant reaction to Tyndall's 1874 Belfast Address seems to have been much louder than Catholic reaction (though both denominations were hostile), when Tyndall wrote an apologia for the speech a few months later he included a rather unconvincing *post facto* justification of it as an attack on the Irish Catholic hierarchy in retaliation for their supposed refusal to allow the teaching of science at the Catholic University.[8]

Richard Jarrell has written of the increasing involvement of the British government with Irish science during the nineteenth century,[9] and this created a second strand within Irish science based in institutions largely created by the Ascendancy. These institutions were the Dublin Museum of Science and Art (now known as the National Museum), the Botanic Gardens at Glasnevin north of Dublin, the Royal College of Science for Ireland and the Irish branch of the Geological Survey, all of which were being administered by the South Kensington Department of Science and Art by 1899. Many of these scientists were English or Scots, and it is the conflicting perceptions of the Irishness of Irish science among these 'administration scientists' which forms the basis of the case studies below.

The third strand of Irish science is that of the majority Catholic and nationalist tradition, in the nineteenth century mainly located in the two overtly Catholic institutions of higher education – St Patrick's College, Maynooth, and the Catholic University founded by Cardinal Newman which later became University College Dublin. The mutual refusal of the Catholic hierarchy and successive British governments to come to terms on the Irish University Question denied many Catholics the chance of higher education, but the National University of Ireland took off with some success from 1908.

On the whole these three strands coexisted fairly peacefully. The Royal Irish Academy may have been partly responsible for this: originally very much an Ascendancy body, around 1900 it tightened up its membership qualifications and, more importantly, decided to retire a fifth of its council every year, greatly speeding up the rate of turnover. This immediately made the council less of a Trinity College clique, and indeed after 1910 or so all three of the groups of scientists had roughly equal representation on the academy's council. By 1925, 35 members of the academy held or had held posts at TCD, 33 were affiliated similarly to the NUI, 8 were at Queen's, and 35 were at other teaching bodies and institutions like the National Museum or the Royal College of Science. Whether or not the reforms around 1900 were intended to facilitate access to the academy from all parts of Irish science, they certainly did have this effect.[10]

This is probably why the academy survived the only serious attempt to introduce new divisions among Irish scholars, an attempted secession from it by nationalists. In April 1921, Timothy Corcoran, the Professor of Education at UCD, organised the setting-up of a National Academy of Ireland, partly as a reaction to the RIA's refusing to readmit Eoin MacNeill as a member – he had been expelled after the 1916 Rising. The National Academy's initial steering committee of eleven included nine professors of the NUI, and five of them were scientists. It immediately started lobbying the revolutionary Dáil government, which looked increasingly likely to become the real government, for recognition as the representative body of Irish learning. The National Academy's foundation meeting was held on 19 May 1922; it had 105 members, of whom about a third were teaching at NUI, none at Trinity College and one, R.M. Henry, at Queen's. Most of NUI's scientific staff joined the National Academy; the future for the Royal Irish Academy looked bleak.

The National Academy of Ireland however sank without trace in 1922. The civil war obviously played a part in this; Count Plunkett and Eoin MacNeill, the two members most active in national politics, ended up on opposite sides. The RIA also moved remarkably quickly once the threat became obvious; it reinstated MacNeill forthwith and produced a propaganda leaflet about its previous services to Celtic studies and its confident expectation that it would continue to be a national institution. Michael Collins' Provisional Government, which was now in power, announced that it would continue state payments to the RIA 'for the time being', and it and its successors have never stopped doing so.

It is interesting that scientists in Finland suffered an organisational split between the Swedish-speakers and Finnish-speakers in 1908, echoing an earlier split among Finnish biologists in 1896 and preceding a further split among Finnish chemists in 1918. Finland, like Ireland, gained independence in the aftermath of the First World War; it too promptly suffered a civil war, much worse than the Irish one; it too had – and still has – a fundamental cultural division, with a Swedish-speaking population roughly equal in proportion to that of the Protestant population of the newly independent Irish Free State.[11]

An important difference between the Irish and Finnish situations was – and is – that in Ireland, religion is used as a signifier to separate two largely English-speaking cultural traditions, while in Finland, language is used as a signifier to separate two largely Lutheran traditions. Since the Irish language was never going to be used as a major vehicle for scholarly communication even by its enthusiasts, the RIA's use of English did not exclude Irish nationalists in the way that the older Finnish societies' use of Swedish did exclude Finnish nationalists. Protestant and Catholic scholars participated in the same discourse; Swedish speakers and Finnish speakers could not.[12] It is also worth noting that the Finnish splits mainly took place before independence, and the failure of the National Irish Academy was perhaps due to its appearing too late in the day.

A LITTLE BIBLIOMETRY

One might have expected science to undergo something of a retreat in Ireland at the start of this century as the supposedly non-scientific and Romantic Irish literary revival and the philosophy of Irish-Ireland gathered in strength. However a citation analysis of three recent survey articles dealing with different aspects of the natural environment of Ireland suggests that, at least in certain areas of science, the 1900–1914 period saw a surge of activity, damped by the outbreak of war, which was not again reached for decades. In all three cases the peak is mainly due to scientists in the direct employ of the government. The graphs below show the date of publication of articles and books cited in each of the survey articles.[13]

The surge of publications on collembola (spring-tailed land invertebrates) shown in the first graph is single-handedly due to the efforts of G.H. Carpenter (1865–1939), Assistant Naturalist in the Natural History Department of the National Museum from 1888, and from 1904 to 1921 Professor of Zoology in the Royal College of Science for Ireland. Carpenter published 32 of the 40 articles cited by Bolger published between those

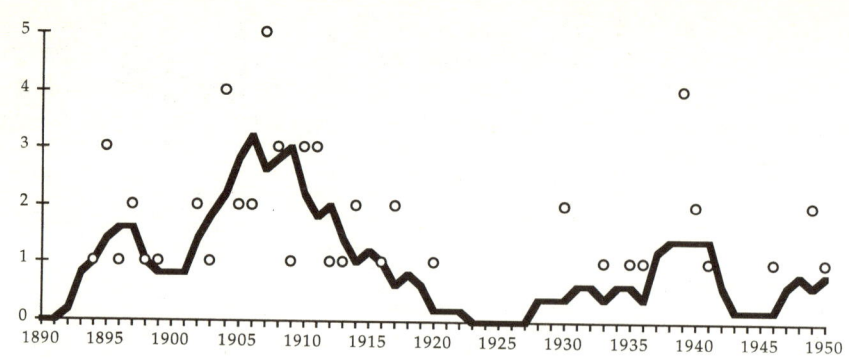

Graph 1. Citations in T. Bolger, 'The Collembola of Ireland'.

○ Articles published in each year — 5-year average of articles published

dates. Carpenter's volume (and indeed quality) of output on entomology was not to be equalled until the mid-1970s.

The contributors to the peak in Irish vegetation studies shown in the second graph were more numerous, though again a substantial contribution was made by one person, Robert Lloyd Praeger, an employee of the National Library (which was also one of the Science and Art Institutions) who was deeply involved in the natural history movement and had competed unsuccessfully with Carpenter for the assistant naturalist post in the National Museum in 1888. Praeger was nonetheless easily the most influential Irish biologist of his time.[14]

The surge of publications on the Rockall Trough, off the western coast of Ireland, was due to a sudden injection of state funding into Irish marine biology. In 1900, the newly created Department of Agriculture and Technical Instruction for Ireland took over the Inspectors of Irish Fisheries, who had existed for decades in order to make and enforce fisheries by-laws, and equipped them with a scientific staff and a steamer for marine

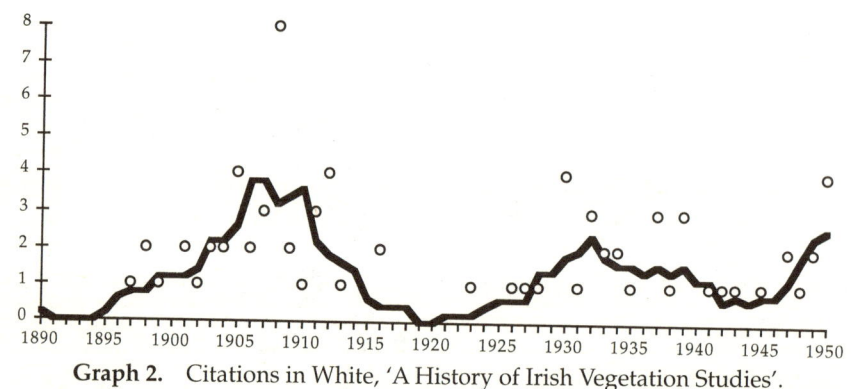

Graph 2. Citations in White, 'A History of Irish Vegetation Studies'.

○ Articles published in each year — 5-year average of articles published

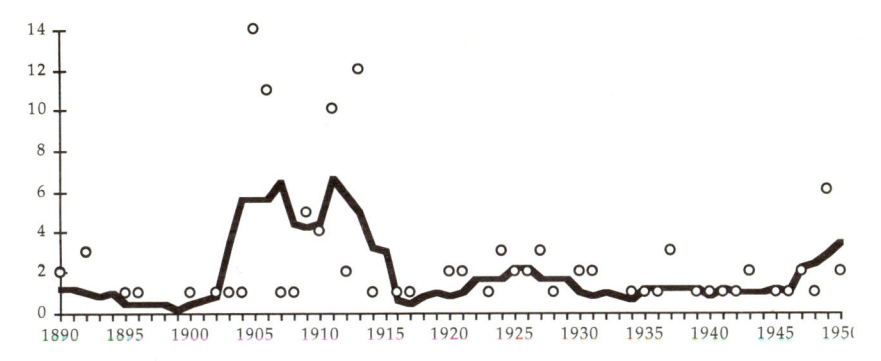

Graph 3. Citations in Mauchline, 'Bibliography of the Rockall Trough'.

○ Articles published in each year   ——— 5-year average of articles published

patrols and research. This steamer, the *Helga*, achieved lasting fame in Irish history when it was used as a gunboat during the 1916 Rising to shell the rebels out of the GPO. The head of the scientific staff of what became known as DATI's Fisheries Branch was Ernest W.L. Holt, an old Etonian who had been invalided out of the army after the Burma campaign in 1887, and had dedicated himself to marine biology and participated in some of the earlier surveys of Irish fishing grounds in the 1890s.[15] As well as surveying the species and abundances of fish off the Irish coast, Holt and his team managed to find many new species of marine crustacea, of which more presently.

Before looking at the story of how the scientists of the administration reacted to the political currents in which they were immersed, it seems appropriate to comment a little further about the usefulness and limitations of citation analysis as a tool for the historian. It is not a perfect diagnostic tool. One recent handbook to the discipline warns that:

> ... interesting or pioneering works do not count for any more in statistics of publication than the great mass of mediocre or indifferent works ... Part of the problem with primitive publication statistics is, of course, that they measure something different to what the historian and theorist of science are really interested in ... We can provisionally conclude that quantitative historiography, based on counting scientists, publications or discoveries, is encumbered with considerable methodological defects and an inbuilt bias.[16]

In any case, when we look at publication rates in the 1860s and the 1960s, or even the 1890s and the 1930s, we are not comparing like with like. As two other researchers point out,

> Through the Victorian period the *conditions* on which an individual participated in the British scientific enterprise were far more flexible than they became later on ... The British scientific enterprise proceeded as it did on the top-most intellectual level only because of the varied involvements of very large numbers of individuals whose participation was on *their* terms, rather than on the terms of a nascent scientific professionalism.[17]

This caveat applies to the first two graphs presented here; many of the articles included in the peak years are quite short notes in the *Irish Naturalist*, a journal run by Praeger and Carpenter, each of which is counted as equal in weight to the much longer articles which would might be expected of a scientist in more recent decades. But although the height of the peaks may be exaggerated, there can be little doubt that they represent genuine surges in activity.

The other potential weakness of all three graphs is that they represent works cited in recent review articles by contemporary scientists in the field, and it is therefore possible that works which might be of interest to the historian could be bypassed by the invertebrate zoologist/botanist/oceanographer of the late twentieth century. There is no real way around this problem other than to suggest that, to a certain extent, it cancels out the first problem mentioned; that in fact a contemporary professional is more likely to discount irrelevant and trivial works and that as a result we are more likely to compare like with like.

ADMINISTRATION SCIENTISTS AND NATIONALISM, 1904–1920

The Inspectorate of Irish Fisheries was just one of the scientific functions which the Department of Agriculture and Technical Instruction acquired in 1900. It also took over from South Kensington all the scientific institutions which the government had acquired in the previous century: the National Museum, the Botanic Gardens, the Royal College of Science and from 1905 the Geological Survey as well. The devolution of control of state science in Ireland to an Irish-based government department, one which even had a certain amount of democratic input, led to the potential for a conflict of loyalties among the English scientists who were now working for the greater scientific good of Ireland.

E.W.L. Holt, mentioned above, was one of these. Another, in charge of the Natural History Department of the National Museum from 1887 to 1921, was Robert F. Scharff (1858–1934), born in Leeds of German parentage. Scharff was a zoologist of some note, specialising in what would now be called biogeography, the study of the geographical distributions of different species. His research led him to propose a number of land-bridges, now sunken, which had previously connected land areas now separated by seas or even oceans.[18] Finally, the head of the Geological Survey from the time of its transfer to Irish control in 1905 till his death in 1924 was Grenville Cole 1859–1924), a Londoner who had been the Professor of Geology in the Royal College of Science since 1890.

How Irish were the scientists of the administration? To what extent did their location in Ireland determine the kind of science they were doing, and the way they thought about it? Most of them, particularly those at senior levels, were English by birth; but their correspondence shows that some individual English scientists working in Ireland were prepared to press the claims of Irish science in order to strengthen their own positions. In the three episodes which I examine, the issue is the same: whether uniquely Irish specimens of marine crustacea, trilobites or foraminifera should be preserved in an appropriate Irish institutional setting or in a metropolitan, British base.

THE STATELESS CRUSTACEA

The first of my three case-histories of Irish nationalism in science involves R.F. Scharff of the National Museum and E.W.L. Holt of the Fisheries Branch. In 1904 and 1905, Holt was debating what should be done with the increasing number of specimens of marine crustacea which he and his colleagues were collecting from the seas off Ireland. A very large number of these were creatures which had not been previously described by scientists; a paper by G.P. Farran in the 1902–03 Fisheries Report lists thirteen new species (although Farran subsequently withdrew his claims to several of these), and a paper by Holt and W.M. Tattersall in the same report lists thirteen more.[19] In particular Holt was concerned about the fates of the type specimens (those specimens from which the official description of a species is drawn) of the new species being discovered. He shared his worries with W.T. Calman, the Assistant in charge of the crustacea collections in the South Kensington Natural History Museum, in October 1904:[20]

> I am most anxious not to have the question of disposal of material raised. If it is raised I may some day be compelled to hand over types of new species to our own museum, where they would be useless.[21]

> Of course I want the types to remain in the British Museum, which is the proper place for them, but I don't want the fact advertised when they are solitary specimens.[22]

Holt's dilemma was clear; his instinct was that the British Museum should be the repository of type specimens, not just for Britain but for the rest of the Empire if not the world. However, he was aware of the rise of nationalism in Ireland, and was alert to the prospect that Scharff might demand that the type specimens of new Irish species be deposited in Ireland's National Museum. Indeed, a year later, this came to pass:

> Scharff, Ph.D., is organising a campaign against me for presenting Irish types to the B.M. He daren't move in the matter personally, so he is putting in some outsider to call attention of Dept to this scandalous transaction and has kindly offered, if I will promise never to repeat the offence, to ward off the attack. I have told him that my hand has generally been strong enough to keep my head from worse dangers than this particular feud is likely to bring on me, and have suggested that his friends might be more profitably engaged in agitating to get his place made into a more fit receptacle for valuable specimens. The ways of the Irish patriot of Teutonic extraction are truly edifying.[23]

If Holt found a new species in British waters, doing research paid for by the British tax-payer, he intended to send the type specimens of the new species to the British Museum. Scharff, on the other hand, argued that they were Irish crustacea, discovered by a branch of the same Irish department which ran the National Museum, and that the type-specimens belonged as of right to Dublin.

It seems that Scharff lost the argument on this occasion; the National Museum of Ireland's catalogue of marine crustacea includes type specimens of only seven species, two of which were discovered in 1896 and the others after 1910.[24] Various specimens of the species which were discovered by Holt and his colleagues between 1900 and 1910 are listed but none is described as a type specimen, so we must assume that the types were indeed sent to, and remain in, the British Museum.

THE TYRONE TRILOBITE

A very similar dispute arose between geologists in Ireland and Britain in 1915, this time over a number of fossils including the type-specimen of a trilobite originally discovered in County Tyrone around 1840.[25] One of Grenville Cole's distant predecessors in charge of Irish geology was Colonel Joseph E. Portlock, an officer of the Ordnance Survey. Portlock suffered generally from an inability to see the wood for the trees; he completed a survey of half of Tyrone in the time he had been allowed for surveying the whole of County Derry, operating from the army barracks in North Queen Street, Belfast. Not surprisingly he was eventually transferred to other duties in Corfu. However, Portlock had managed to amass a substantial collection of fossils – unlike his immediate superiors he had realised the value of palaeontology for dating strata.[26] Some of his fossils were transferred to London from the Museum of Irish Industry collections at some point before 1870.

In June 1915, the Board of Education in England received a request from T.P. Gill, the Secretary of DATI, that Portlock's fossils be returned. Attached was a memorandum from Cole, putting the case for returning the Irish fossils to their former home:

> In spite of its peculiar significance in the history of geological research in Ireland, [Portlock's collection] was at some date about 1870 transferred to the Museum of Practical Geology in London. The fossils are there 'amalgamated with other collections', and are distributed throughout the general collection of fossils of the British Isles. Numerous specimens described by Portlock appear in 'A catalogue of Cambrian and Silurian fossils in the Museum of Practical Geology' issued for H.M. Stationery Office in 1878.
>
> I understand that certain specimens have escaped from this collection into the teaching-collection of the Imperial College of Science . . . and that Portlock's type-specimen of a trilobite, *Nuttainia hibernica,* from Tyrone, has been re-discovered with the original Ordnance Survey label attached to it. Now . . . it would seem highly desirable if the Department could approach the Board of Education in London, with a view to the return of the Portlock fossils to the collections of which they originally formed a part. It must be remembered

that their removal may have seemed natural when the Geological Survey was administered from London as a whole; but this is the only case, so far as I am aware, when a distinctively Irish collection was so transferred.[27]

This caused some consternation in the British geological survey; anxious memos flew for several weeks between its director, A. Strachan, and F.G. Ogilvie, Permanent Secretary to the Board of Education. They eventually concluded that, first, it would be completely impractical to separate out the fossils from any one particular source in the Museum of Practical Geology's collection; second, in any case the collection had probably been divided up long before 1864; and finally:

> The escape of the trilobite to which Professor Cole refers dates from a time when the Collections at Jermyn Street were used for teaching purposes in the Royal School of Mines then carried on in the same building. These Collections have not been used in this way for thirty-five years.[28]

This last barb was particularly intended to forestall any further efforts on Cole's part; Ogilvie, while going through the records, had discovered with glee that Cole himself had been employed as a demonstrator in Jermyn Street at the time of the trilobite's 'escape'.[29] 'The more one looks into it,' he commented to Strachan, 'the more does Prof. Grenville Cole's memorandum suggest the mode of presentation adopted in the Notes which the German Foreign Office sends to Washington!'

There is a postscript to this story. Two decades later in 1936 the National Museum of Ireland and the Irish Geological Survey discussed the possibility of returning the Portlock fossils to their mother country; the Geological Survey informed the museum that Cole's attempt to do so had been unsuccessful, and that furthermore all the Portlock fossils had been collected in the six counties which remain under British rule. They remain in London to this day.[30]

The Wright Foraminifera

My third case study concerns the foraminifera collection of Joseph Wright (1834–1923), which actually brought Cole and Scharff into conflict with each other. This has been reconstructed from the archives of the South Kensington Natural History Museum on the one hand[31] and the National Museum[32] on the other. Joseph Wright was a Quaker from Cork, who made a successful career as a grocer in May Street in Belfast. Wright had become one of the world's leading experts on foraminifera, and was consulted by Canadian and Australian provincial governments. Robert Lloyd Praeger reminisced of him affectionately:

> A more kindly enthusiast than Joseph Wright never lived. I remember one occasion on which his self-restraint and benevolence were put to a severe test. In a dredging sent to him from – I forget where – he discovered a single specimen of remarkable novelty – the type of a new genus of Foraminifera. He

> mounted it temporarily on a slide – neglecting to put on a protective cover-glass, for he was a careless manipulator – and at a conversazione of the Belfast Naturalists' Field Club held immediately afterwards he showed it to J.H. Davies and others. Davies was a fellow-quaker, an ardent bryologist, a man of singular courtesy, a neat and skilful microscopical expert. Seeing that the slide was dusty, and not noticing the absence of the usual cover-glass, Davies leisurely produced a silk pocket-handkerchief and, before the horrified eyes of the owner, in a moment ground the specimen to powder! But Wright's self-restraint stood even that test. He gasped, and his face went white; but he uttered no word of reproach.[33]

By 1920, however, Wright was well into his eighties and in need of constant care. His relatives decided that the only thing to do was to sell Wright's lifetime collection of foraminifera, and approached the British Museum in February 1920 through William Swanston, another Belfast naturalist.[34] Grenville Cole had somehow heard that the approach was being made, and on 7 July 1920 wrote to Harmer, the British Museum's Keeper of Natural History:

> I hear that you are considering the acquisition of the Wright collection of foraminifera for the British museum. I presume that some sample has been sent to you. Wright himself was a wonderfully careful manipulator, and some of his type-slides are veritable gems.[35] It is very hard to value what are practically unique things, each specimen being selected by an expert, and I have been asked by old friends in Belfast if I can give a fair idea. It strikes me that you may be able to make an offer, since it is a bargain direct between scientific men, and it is so difficult to asses the personal selection and critical knowledge that was involved. 1725 slides at 1/6 makes a good sum; I have no idea what the family would expect. The whole, if I judge Wright's work aright (I love the old man, but must not be biased [sic] that way) is worth £150 to the nation. Is such a suggestion at all improper at the present time? The leisure of a lifetime must be in the collection, stored up for future workers. [inserted: Please do not send my proposal to Belfast!] If you have any views, I can hand them on or not, exactly as you wish. Luckily, I have no notion whatever of the expectations of the Swanston-Wright combine.[36]

Cole added in a revealing postscript:

> I am glad that the collection should find a home in the British Federation. Dublin is no secure place for Commonwealth treasures now.

The British Museum was feeling a post-war spending pinch, but managed to contact an amateur English geologist, Edward Heron-Allen, who knew and admired Wright, and was also an expert on foraminifera.[37] Heron-Allen was very keen to make an offer for the whole collection:

> Of course Joseph Wright's collections *must* come to the B.M. if not direct, then through me. I will give him £200 for the slides and library if you can arrange it with him – I can't pay for them until the end of the year.[38]

Wright's wife and daughter, apparently in the mistaken belief that Heron-Allen's intention was to keep the foraminifera in his own collection, felt that they would rather sell to a museum than a private collector. Accordingly in late August 1920 Swanston wrote on their behalf to R.F. Scharff, who had been the Acting Director of the whole of the National Museum since Count Plunkett had been sacked in the aftermath of the Easter Rising:

> Mrs Wright and daughters, – with whom I sincerely concur, – would much prefer that the Collection should find a resting-place in an Institution available to the public, hence my addressing you with the view of you being able to secure it for your museum. And may I state, that, the Collection being largely derived from Irish localities it would, in my opinion, be very suitably and usefully housed in Dublin.[39]

After a short delay while Swanston had Wright's catalogue returned from South Kensington in order to send it off again to Dublin, Scharff approached T.P. Gill, still Secretary of the whole Department of Agriculture and so in control of expenditure, to ask for a special grant of £200 to buy the collection. Gill asked for a reference for the grant; by the end of October, Scharff had got a suitable reference from G.H. Carpenter of the Royal College of Science, insisting that the Wright collection was unique, and that it should be preserved for Ireland. Equipped with Carpenter's reference, Gill approved the grant, and Scharff wrote to Swanston on 1 November with the good news, enclosing pre-addressed labels to ensure speedy delivery.[40] Heron-Allen was naturally disappointed:

> Rather a blow! But it is my own fault; of course I ought to have given *you* the £200 to buy the Collection, & you could have 'lent' it to me for life, so that it would have been catalogued and ranged with all the other collections. But that's *that*![41]

THE MOVING METROPOLIS

The political background against which these disputes took place changed considerably between 1903 and 1920. Scharff's attempt to claim Ireland's newly found crustacea for the National Museum coincided with the devolution crisis of 1904–05, when the Earl of Dunraven, with the active support of the most senior Irish civil servant, Sir Antony MacDonnell, attempted to persuade the unionist government to adopt a scheme which fell far enough short of Home Rule to leave nationalists unsatisfied but was sufficiently innovative to infuriate unionists. It is interesting to compare the worried tone of Holt's letters to Calman the month after the devolution crisis first broke in September 1904 with his more usual bombast the following year, some time after the Chief Secretary, George Wyndham, had brought the crisis to an end by resigning.[42]

Irish Home Rule made its way onto the statute book in 1914, despite the opposition of the House of Lords and of the greater part of the population of Ulster. However, its implementation was to be delayed until the end of the war which had broken out in August 1914. Unionists and a majority of nationalists agreed to forget their political differences in the common cause of an Allied victory which nobody expected to take four years. By 1915, the initial enthusiasm was beginning to wear off Ireland's war effort, and the discontent which culminated in the following year's Rising was increasing. The British attitude seems to have been to reject the warning signs as mere whingeing; had the Irish not been promised that they could have Home Rule, perhaps with some special provision for Ulster, once the Germans had been dealt with? Cole's rebuff over the Portlock collection seems in keeping with British policy in general; in the midst of a world war, and at a time when traditional Irish nationalism was at a low ebb, there was little political mileage to be gained from trilobites.

The years after the war ended brought the collapse of first traditional Irish nationalism and then British rule in most of Ireland. By the summer of 1920, the 'War of Independence' was about to enter its nastiest phase. It was at this stage that the government authorised the military to carry out reprisals against the civilian population – probably in hindsight the best possible way of encouraging a revival in support for Sinn Féin and the violent independence movement. The political supremacy of the independence movement had been consolidated in the January 1920 local elections; 'by the summer of 1920 the civil authority in many rural areas was effectively subservient to Sinn Féin'.[43] It must have seemed certain to Cole, Gill and Scharff that Sinn Féin were the future government of Ireland. Cole was concerned for the availability of Wright's collection to future generations of students, but also for the future of the relationship between the South Kensington museum and Irish science. Scharff was responsible for building up the museum's collections, and for safeguarding it as a national institution against whatever storms might lie ahead. Gill too wanted to preserve the various parts of the department for which he had worked so hard for whatever form of government was to emerge.

These three episodes are very reminiscent of the recently retold story of the Melbourne meteorites.[44] Two fragments of what was believed to be the largest known meteorite were identified near Melbourne in 1860; the ownership of the fragments was disputed between the National Museum of Victoria and the British Museum, represented respectively by the Irish palaeontologist Frederick McCoy (1823–99), the National Museum's director, and the Austrian Ferdinand Mueller, the Government Botanist and Keeper of the Botanic Gardens. Politicians in Victoria, including the colony's Governor, became involved with the dispute, which ended with the larger fragment being moved to London and the smaller remaining in Victoria.

For Ireland's administration scientists, the location of the metropolis became distinctly uncertain between 1900 and 1922. Scharff would have had great difficulty even in initiating a dispute with South Kensington before the Dublin Museum's administration was transferred to DATI in 1900. But once the Dublin Museum of Science and Art had become the National Museum of Ireland, Scharff could assert that it had a right to Ireland's national heritage, and that he was entitled to deny parts of that heritage to the imperial centre. Holt, also an Englishman working in the Irish scientific enterprise, could assert that specimens in Dublin 'would be useless' and that South Kensington was 'the proper place for them' without needing to defend his statements. Cole wanted to repatriate an important Irish fossil collection in 1915, but five years later did his best to prevent another collection being kept in Ireland. Whether Irish specimens belonged in a British or an Irish museum became a metaphor for Ireland's preferred political status among individual scientists. Scharff and Cole lost the disputes with Holt and the British geologists, at times when Home Rule seemed a long way off; but Scharff was able to gain the Wright foraminifera when Irish independence was imminent and the political balance was shifting in favour of the National Museum of Ireland of which he was the acting director.

The cases of the stateless crustacea, the Tyrone trilobite and the Wright foraminifera demonstrate the effects of the changing Irish political environment on those scientists working within it. Science may perhaps be an international pursuit, but the actions of scientists are affected by issues of national allegiance. Scientists within the Ascendancy or nationalist traditions were instinctively attracted to a particular view of their own role within the British Empire or the Irish Nation as the case might be. Those like Holt, Cole and Scharff who were in the employ of the state in Ireland and were not themselves Irish-born had a more difficult path to negotiate between the conflicting claims of imperial and national science.

NOTES

1. J.W. Foster, 'Natural Science and Irish Culture', *Éire-Ireland* 26 (1991): pp. 92–103.
2. M. Ward, *The Missing Sex, Putting Women into Irish History* (Dublin, 1991).
3. G. Basalla, 'The Spread of Western Science', *Science* 156 (1967): pp. 611–22.
4. *Scientific Colonialism: A Cross Cultural Comparison*, ed. Nathan Reingold and Marc Rothenberg (Washington, 1987) and *Science and Empires: Historical Studies about Scientific Development and European Expansion*, ed. P. Petitjean, C. Jami and A.-M. Moulin (Dordrecht, 1992) based on the proceedings of conferences held in Melbourne in 1981 and Paris in 1990 respectively.
5. R. Macleod, 'On Visiting the "Moving Metropolis": Reflections on the Architecture of Imperial Science' in Reingold and Rothenberg pp. 217–250.
6. I have found very helpful the three strand model proposed in V. V. Krishna, 'The Colonial "Model" and the Emergence of National Science in India: 1876–1920' in Petitjean et al pp. 57–72.
7. Beckett, *The Making of Modern Ireland 1603–1923* (London, 1966) and Foster, *Modern Ireland, 1600–1972* (London, 1989).
8. For both the 'Belfast Address' and the 'Apologia' see Tyndall, *Fragments of Science, vol. II* (London, 1879).

9. Richard Jarrell, 'The Department of Science and Art and control of Irish Science, 1853–1905', *Irish Historical Studies* 23 (1983): pp. 330–347.
10. See R.B. McDowell, 'The Main Narrative' in T. Ó Raifeartaigh (ed.): *The Royal Irish Academy: A Bicentennial History, 1785–1985*, (Dublin, 1985).
11. My attention was first drawn to the similarities between Ireland and Finland by Jonas Jørstad, 'Nations once again', in *Revolution? Ireland 1917–1923*, ed. David Fitzpatrick (Dublin, 1990). For the splits in the Academy of Science/Society of Science and among Finnish biologists, see R. Collander, *The History of Botany in Finland, 1828–1918* (Helsinki, 1965) p. 107–111; for the split among Finnish chemists see T. Enkvist, *The History of Chemistry in Finland, 1828–1918* (Helsinki, 1972).
12. That the language issue in Finland was a matter of personal political choice rather than family background is demonstrated by the story of the Neovius family, which produced a number of notable mathematicians throughout the nineteenth century. Edvard Rudolf Neovius was the Professor of Mathematics in Helsinki from 1883 until he was made the Minister for Finance in 1900. The Finnish government of 1900–1905 became deeply unpopular with nationalists who perceived it as too subservient to the Russians, and when it collapsed Neovius had to emigrate to Denmark. His two brothers, Lars Theodor and Otto Wilhelm, both prominent mathematics teachers, changed their surname to the more patriotic sounding Nevanlinna in 1906. See G. Elfving, *The History of Mathematics in Finland, 1828–1918* (Helsinki, 1981) pp. 108–112.
13. The articles are T. Bolger, 'The Collembola of Ireland: A Checklist and Bibliography', *Proceedings of the Royal Irish Academy B* 86 (1986): pp. 183–218; J. White, 'A History of Irish Vegetation Studies', *Royal Dublin Society Journal of Life Sciences* 3 (1982): pp. 15–42, and J. Mauchline, D.J. Ellett, J.D. Gage, J.D.M. Gordon, and E.J.W. Jones, 'A Bibliography of the Rockall Trench', *Proceedings of the Royal Society of Edinburgh B* 88 (1986): pp. 319–354.
14. Apart from Seán Lysaght's as yet unpublished Ph. D. thesis, *Robert Lloyd Praeger and the Culture of Science in Ireland: 1865–1953* (NUI, 1994), sources on Praeger include Timothy Collins, *Floreat Hibernia: A Bio-Bibliography of Robert Lloyd Praeger* (Dublin, 1985), and Praeger's own *The Way that I Went: An Irishman in Ireland* (Dublin, 1937) and *A Populous Solitude* (Dublin, 1941).
15. Biographical details about Holt, Scharff and Cole are based on the relevant entries in Robert Lloyd Praeger, *Some Irish Naturalists* (Dundalk, 1949), with some necessary corrections based on O'Riordan, *The Natural History Museum* (Dublin, c. 1983) pp. 28–32 and 60.
16. Helge Kragh, *An Introduction to the Historiography of Science* (Cambridge, 1987) p. 186, 187, 189.
17. S. Shapin and A. Thackray, 'Prosopography as a research tool in history of science: the British scientific community 1700–1900', *History of Science*, 12 (1974): pp. 1–28, p. 12.
18. See Scharff, *European Animals* (London, 1907).
19. The papers are appendices to the *Irish Fisheries Report, 1902–03, Part II* (1905) Cd.2535: G.P. Farran, 'Report on the Copepoda of the Atlantic Slope off Counties Mayo and Galway', Appendix II, pp. 23–52, and E.W.L. Holt and W.M. Tattersall, 'Schizopodous Crustacea from the North-East Atlantic Slope', Appendix IV.i, pp. 99–152.
20. All the following letters are to be found in the Crustacea Correspondence in South Kensington.
21. Holt to Calman, 4 October 1904.
22. Holt to Calman, 12 October 1904.
23. Holt to Calman, 12 October 1905.

24. C.E. O'Riordan, *A Catalogue of the Collection of Irish Marine Crustacea in the National Museum of Ireland* (Dublin, c. 1969).
25. The material for this case-study is taken entirely from the British Geological Survey archive file GSM 2/409.
26. Herries-Davies, *Sheets of Many Colours*,(Dublin, 1983), pp. 95–106.
27. Undated paper in GSM 2/409, dated receipt 15 June 1915.
28. Typed memo from Ogilvie in GSM 2/409, dated 27 July 1916.
29. Hand-written memo from Ogilvie to Strachan in GSM 2/409, dated 16 July 1915.
30. Natural History Museum, Incoming Correspondence.
31. South Kensington Archives, Keeper's Correspondence, 1920.
32. Natural History Museum, Official Correspondence and Incoming Correspondence drawers.
33. Praeger, *Way that I Went* , pp. 166–167.
34. Keeper's Correspondence 1920/72, Swanston to Harmer, 10 February 1920.
35. Rather different from Praeger's characterisation above! Perhaps Wright had learnt from experience to be more careful.
36. Keeper's Correspondence 1920/71, Cole to Harmer, 7 July 1920.
37. He also wrote science fiction under the pseudonym Christopher Blayre. See J. Clute and P. Nicholls, *The Encyclopedia of Science Fiction* (London, 1993).
38. Keeper's Correspondence 1920/72, Heron-Allen to Harmer, 29 July 1920.
39. Swanston to Scharff, 16 August 1920 (Natural History Museum Archives, Scharff correspondence).
40. Natural History Museum Archives, Official Correspondence drawer, Wright Collection folder.
41. Keeper's Correspondence 1920/72, Heron-Allen to Harmer, 11 November 1920.
42. O'Halpin, *The Decline of the Union* (Dublin, 1987), pp. 44–51.
43. R.F. Foster, *Modern Ireland* p. 497.
44. A.M. Lucas, P.J. Lucas, T.A. Darragh & S. Maroske, 'Colonial pride and metropolitan expectations: the British Museum and Melbourne's meteorites', *British Journal for the History of Science* 27 (1994): pp. 65–87.

Chapter 5

James Ryan (c.1770–1847) and the Problems of Introducing Irish 'New Technology' to British Mines in the Early Nineteenth Century

Hugh Torrens

INTRODUCTION

The life and work of the Irish-born mining surveyor and engineer James Ryan throws light on the problems encountered by early nineteenth-century inventors trying to interest the commercial world in their work. Ryan invented a boring device which for the first time recovered cores and so gave superior information about underlying rock strata. This invention could also be used to ventilate mines. His work enjoyed only partial success in Ireland, and he encountered much opposition when he moved to England later in his career. Ryan's experiences show how even important inventions can be rejected on the grounds of short-term expediency. It is also apparent that his Irish origins generated a certain amount of prejudice against him among the already-suspicious English mine owners.

RYAN'S IRISH ORIGINS

From the commonplaceness of his name, Ryan's early history must remain unknown. He was certainly born in Ireland in about 1770 (from his age at death). Conceivably a Tipperary land surveyor and grocer of the same name, recorded as active from 1750 might be related.[1] A later claim that Ryan hailed from Donegal relates only to his work there in 1814 (see below).[2]

Ryan later cryptically noted that 'whilst endeavouring, since 1792, to gain a living in Britain, I adopted mining as a calling'.[3] This introduces a second problem for the historian, namely the papyrophobic nature of much mining activity. Ryan noted in one of the few pieces of autobiography that he wrote that he had 'from a youth, a powerful predilection for the study of minerals and their various geological character and position'.[4] This brought him his first post in Ireland in 1800, a date confirmed by his 1831 note that he had been '30 years as a miner'.[5] Ryan later acknowledged that he then 'became noticed by some partial friends who had attached great importance to the systematic arrangement of my ideas'. Among these was the celebrated Irish scholar Richard Kirwan, who had published *Observations on Coal Mines* in 1789.[6] Since Ryan's birthplace is unknown, whatever schooling he received is equally uncertain. Ryan's own copy of Samuel Cunn's London edition of *The Elements of Geometry* dated 1754 which survives, may be a relic of his school days.[7]

Ryan's first mining employment was as 'mineral surveyor' working for the Grand Canal Company of Ireland (GCI) from 1800.[8] This company soon started a new and expensive mining venture in 1803, to build up trade on the Canal, by leasing some collieries near Castlecomer.[9] By April 1804 Ryan was based at these Doonane collieries, north-east of Castlecomer on the borders of the then counties of Kilkenny and Queen's County (now Laois). This was in an area of considerable technical innovation in an Irish context, both for canals and collieries. A new canal from Monastereven to Castle Comer had been proposed in 1800, with a branch to 'the coalpits'[10] and at these Doonane pits both the first Newcomen steam engine in Ireland had been erected, in 1740, as well as the first separately condensing Boulton and Watt steam engine, in 1782–83.[11] But the output of the Doonane collieries had not been able to justify the high and, with yearly premium payments, recurring costs of this last innovation and they had gone bankrupt in 1791.

GCI archives show Ryan was busy exploring for coal by boring at Doonane late in April 1804.[12] He was soon in dispute with his English superior, Israel Rhodes, perhaps because Rhodes 'does not seem to have had any mining experience'.[13] The GCI Board supported Ryan over Rhodes in July 1804 and this may have precipitated Rhodes' return to England in November 1804. In the following months Ryan was quarrelling with David Aher (1780–1841) who replaced Rhodes.[14] The matter in dispute was whether the GCI should stop exploring and start working some of their coal pits. Otherwise Ryan could see the expense would accumulate and ruin the company to 'the disgrace' of those involved earlier.[15]

By this date Ryan was also busy in North Wales. He had been employed here by the coal owner Edward Roscoe at a Bagilt colliery and nearby by Sir Thomas Mostyn, Bart (died 1831). Sometime later in 1804 he was also 'employed by Rigby & Hancock at Llanwarder Colliery, near Bagilt' again in North Wales.[16]

RYAN'S CORE-BORING INVENTION

Ryan's most significant invention, his new boring apparatus, is first heard of early in 1804. At a meeting of the Dublin Society on 19 April 1804 with antiquary General Charles Vallancey (1721–1812) of the Royal Engineers in the Chair (whose help Ryan later acknowledged[17]), it was noted that 'Mr James Ryan, having laid before the Society a core, brought up by a new borer of his invention, which appears to them to have great merit, and that Mr Ryan is deserving of public encouragement.' It was then resolved 'that Mr Ryan be applied to for information, if he will furnish the Society with such a borer of his invention, to sink 40 feet: and what will be the expense of the same'.[18] On 22 November 1804 'Mr James Ryan [again] attended the Society, and informed them that in compliance with their requisition . . . he had made a boring instrument, for which he had obtained a patent, and was ready to furnish the Society with one of six feet diameter, made exactly according to the tenour of the patent.'[19]

Ryan here claimed to have patented his boring apparatus by 22 November 1804 but this clearly only referred to the fact that he had instigated patent proceedings. His English patent was set in motion for him between 24 January 1805 and 12 February 1805 by an attorney, Walter Henry Wyatt of Hatton Garden, London, while Ryan was still in Ireland. He was named as 'engineer to the Undertakers of the Grand Canal', which company made him first a loan of £120 and then 'paid the entire expense' to allow him to take out this patent.[20] The patent was granted on 28 February 1805,[21] and included a plate with detailed drawings of all the comprehensive elements of his new equipment.[22]

Details of the patent were circulated in both literary and technical literature in England. The first commented that 'as the soil to be bored is of various consistencies in different places, and, at different depths, often times in the same place, Mr Ryan has contrived various bits to be attached to the boring-rod' and concluded, 'by some of the above-mentioned tools, cores or solid cylinders of the soil may be extracted, from one to twenty inches in length, and from one to twenty inches in diameter, by which the dip of the different strata met with in boring may be ascertained, as well as the nature of minerals and other substances which occur.'[23] The technical press also commented favourably that Ryan's patent provided a model, that 'fairly and distinctly' described his invention, unlike so many others which were often 'exhibited to conceal.'[24] The patent was also noticed in German.[25]

The first to publish his observations on Ryan's novel invention was Richard Lovell Edgeworth (1744–1817) of Edgeworthstown, Co. Longford, long a member of the Lunar Society and now an enthusiastic supporter of Ryan's endeavours, as Ryan again acknowledged.[26] Edgeworth wrote in 1806:

> I take this opportunity of mentioning a trial that I lately made of Mr Ryan's patent Boring Machine, for exploring the strata of mineral countries. This

machine acts like the surgical *trepan*, and cuts a circular hole, leaving a *core* in the middle which is drawn up from time to time by a pair of self-closing tongs. I found that this gentleman's machine, from his want of knowledge how to push his own invention, had not obtained due credit; I therefore invited him to try his machine at my house, that I might witness the result of the experiment. Two men, relieved from time to time, cut a truly circular hole, five inches and a quarter in diameter, through a block of hard limestone, leaving a *core* a little taper of four inches and a half diameter and six and a three quarters in length, which core is now in my possession. It is as true and as smooth as if turned and polished in a lathe, and the under surface shews exactly the fracture by which it was detached from the block at bottom. By this contrivance mines may be ventilated at small expense, the specimens of strata that are bored through may be brought up whole and unmixed, no deception can take place; and not only the dip, but the fracture, lap, and accidents to which each stratum is liable, may be determined at any depth. True vertical and horizontal sections may be previously obtained of any spot where it is proposed to sink shafts; and the subterraneous topography of a whole district may be laid down upon a map with confidence before any great expense is hazarded on mere speculation.[27]

Comments agreeing with Edgeworth soon followed: 'We wish [Ryan's invention] more success than it has hitherto met with; for it would be found extremely useful in ascertaining the quality of minerals and other substances, as well as the dip of the strata; at the same time that it is applicable in sinking wells.'[28]

At least two others had tried out Ryan's invention in Ireland, for mineral prospecting. Both wrote letters dated 1806 which were first used in a printed Testimonial for 'J. Ryan, Mineralogical Surveyor, Ventilator and Director of Mines &c', of circa 1812.[29] The first was Samuel Perry (1765–1829) of Woodroof, near Clonmel, County Tipperary,[30] an Irish barrister and land owner who had graduated from Trinity College Dublin in 1786. He 'made several trials on my estate at Woodroof, in search of coal. I found the Patent Engine of your invention vastly to surpass the old system of boring with the chisels and much more satisfactory: your Engine not only bringing up the Cores entire, but it also shews the layer of each stratum as it goes through; whereas, with respect to the chisel, as it cuts every thing to pulp, nothing ... can be ascertained to satisfaction.' The second, L. [recte John von Prebeton, Count] Van Wilmsdorf Richards, was a Hannoverian migrant who married into the Richards family of Mac Mine, Wexford in March 1802.[31] He took the name Richards in May 1802.[32] He thus inherited estates at Rathspeck [Rathaspick], three miles Southwest of Wexford and had 'given over further trial on my estate, because of meeting pudding-stone (a substance supposed to exclude coal) ... [but] I am perfectly satisfied, if any [coal] had been there, by trying with your Patent Engine, I could not have missed it as the cylinders came up perfect, and enabled any person to judge of their properties.'

Edgeworth's first announcement had made it clear that Ryan's invention was already seen to have two main uses, for mineral prospecting (by

The Problems of Introducing Irish to British Mines

cutting cores of the strata penetrated) and for ventilating mines (by boring passages). Richards, by 1806, had also used Ryan's device in successful drainage work, suggesting a third potential use for the invention.

In August 1807, the geologist and polymath John Farey (1766–1826) confirmed that Ryan's invention

> consists in using a cylindrical cutter, something like the surgeon's trepan-instrument, by which a core, or solid and unbroken piece of each stratum, is cut, and by other tools brought vertically to the surface, in the exact position as to the cardinal points, in which it stood in the strata, and thus the quantity and direction of the dip, as well as the exact nature of the strata or measures, are correctly ascertained.[33]

Farey was clearly struck with the novelty of this device by which the crucial dip of strata could be monitored underground.

Ryan had been told by the GCI on 15 April 1805, that 'they will have no further use for his services after 1 May', the GCI having employed Archibald Cochrane (1749–1831), ninth Earl Dundonald to visit and report on the Doonane Collieries on 18 April, with Richard Griffith (1784–1878), later famous as an Irish geologist.[34] Cochrane's long report was very dismissive of Ryan's work, some of whose borings were underneath 'strata known to lay under the Coal' and of his invention.[35] Borehole G which was 35 yards deep had been 'performed with Ryan's large borer . . . in very hard rock . . . and had hitherto cost upwards of £13 per yard, a most enormous price . . . more than double the price at which a Pit could be sunk'. Furthermore Lord Dundonald claimed that he

> does aver that by the old method of boring the strata . . . may be ascertained with the greatest precision and at the same time much less expense than can be done by Mr Ryan's method of boring, which was practised in Scotland 25 years ago and discontinued from its not having been found under many circumstances to answer equally well as the old plan of boring. Lord Dundonald's authority is a Mr Telfer . . . who is now in Dublin.

William Telfer was a mathematical instrument-maker based in Glasgow who had taken out two patents for flax preparation.[36] When James Millar (1762–1827) published his second edition of John Williams' *Natural History of the Mineral Kingdom* in 1810, he confirmed this earlier use in Scotland, but instead noted that it was 'Mr Scott of Ormiston in East Lothian [who] had . . . long employed [such an instrument] for the purpose of extracting a core or cylinder of the coal . . . But I do not know that it was ever applied to rocks in general, or in any other way but to determine the thickness of the coal *when it is discovered by the common apparatus* [emphasis added] . . . Mr Ryan's exclusive privilege may perhaps be considered as perfectly entire.'[37] This confirms that Ryan was the first, at least in Britain, to realise the need for, and to supply a means whereby, cores of strata could be examined when bored through in prospecting for minerals. Ryan's priority is confused by a recent erroneous claim that his invention was developed in the 1820s.[38]

The scale of the potential advance that Ryan had provided can be seen by studying the contemporary alternative 'common apparatus' for boring.[39] This was both old, having then been in use for centuries, and primitive, in that it reduced the rocks penetrated by the borer to fragments by percussion, from which little could be learnt. Ryan's method completely removed the great disadvantage of this old system which was, as Samuel Perry's testimonial of 1806 noted, the way the chisels it used 'pulped' all the strata penetrated and so enormously reduced the quality of any data retrieved. But the old system did have the enormous advantage, in the eyes of conservative British coalowners and of Lord Dundonald, of being cheap. This last 'advantage' allowed it to remain in use into its third century until after the British Coal industry was nationalised in 1945![40]

Ryan leaves Ireland

Such views of the value of 'cheapness' extended to Ireland. Ryan was dismissed from his employment by the GCI in May 1805. He had had a fortnight's notice but on 26 April was told he could now 'stay until 24 May if he likes.' The GCI Board 'thought he would wish to make use of his patent in England', which they could now use in Ireland, having paid for it, 'without the expense of his salary'! Ryan, in a reply to the GCI dated 6 May 1805, noted 'the system of boring, invented by me, has been the only one that the Irish could tell Foreigners that they could [not] be done without.' Ryan claimed 'its expenses must be less than any other as the two [Irish] men that operates [it] does not get one fourth of the wages of the Foreigners. I have proof of sixteen different false reports being made in the Kilkenny & Queen's County Collieries by [old style] boring, since proved by sinking.'[41] All such economies in using Irish workmen in Ireland were however soon found to evaporate when English labour was used in England. The major problem at this date for Irish members of some of the new professions, like mineral surveying, was the lack of openings for them in Ireland.

From this point onwards, Ryan spent most of his time in England. However, he did not cease all contact with Irish mining. On 20 January 1814 he wrote from 'Sudley Mine, Donegal', as Mine agent to the metalliferous mines owned here by the third Earl of Arran, Arthur Saunders Gore (1761–1837). Ryan was 'mine agent to the Earl for nine months' of each year.[42] In the summer of 1814 Ryan was also in Belfast, in contact with the local geologist Dr. James Macdonell (1762–1845).[43]

Ryan and mineral surveying

Edgeworth's note that Ryan's device had two uses is confirmed by the 'schizophrenia' which greeted it when introduced to England. The first was in improving mineral prospecting, where it was more efficient but certainly more expensive. When Ryan arrived in England to further its promotion in the heartland of the Industrial Revolution, most mine owners were interested in costs, not in 'efficiency gains.' So his invention seems to have been hardly put to the use to which members of geologist William Smith's school at least, thought it should have been, namely in prospecting for stratified minerals.

Ryan actively tried to promote this first use in various ways. In April 1807 he wrote to the London-based Board of Agriculture about the use of his invention 'in draining experiments'. In January 1808 his paper 'on his Auger' was referred to its general committee and Davies Giddy (1767–1839) reported on it in July 1808.[44] Sadly the board's further records have not survived but John Farey confirmed that

> in April, 1807, Mr Ryan presented a complete set of his apparatus to the Board of Agriculture in London, and bored a hole of some depth therewith near Kensington, under the inspection of some of its members; the cores or borings therefrom being exhibited to the Board, and lodged with the apparatus in their repository, they voted a pecuniary reward to Mr Ryan. From the apparent importance of this discovery to mining, but to coal-finding in particular, we were induced to wish, to give an accurate description and drawings in this place of Mr Ryan's apparatus . . . [but] under the article MINING we shall endeavour to give them in the further state of perfection, in which practice will doubtless then present the same.[45]

Sadly Farey's problems with the editor of the Rees' *Cyclopaedia*, for which he ceased to write from September 1811, meant that this promise was never fulfilled.[46]

By June 1807 Ryan had also demonstrated his boring apparatus, to the forgotten ancestor of the London Geological Society (founded November 1807), the British Mineralogical Society. Their Secretary was the quaker social reformer William Phillips (1770–1843). His testimonial noted that 'after a careful inspection and examination of the Instrument . . . they are persuaded that it affords results much more satisfactory than any of the methods now in use; since by bringing up the different Strata in solid Cylinders, it supplies the means of a more accurate knowledge of their depth and quality, while it indicates, at the same time, their dip and inclination'.[47]

Later the same year, 1807, Ryan was back in Cumberland, then one of the most technologically advanced areas of coal mining in England. Moreover it provided the great proportion of the coal then imported into Ireland. At Workington the new Isabella Pit was sinking, having started early in 1807. It must have been then that Ryan was given a manuscript section of the strata penetrated during sinking of the pit, down to 168 feet. This Ryan later donated to Trinity College Dublin where it survives.[48] The coal owner here, John Christian Curwen (1756–1828), provided a testimonial on 30 September 1807 that Ryan's boring apparatus had been successfully used there, down to eight fathoms and that he 'had no hesitation in stating it as my opinion, that it will be found highly serviceable to the mining interests of the country'. Earlier the same month Ryan visited the nearby Seaton Iron Works whose manager also provided a highly satisfied testimonial.[49]

Ryan also returned, less successfully, to Whitehaven where he demonstrated his boring apparatus in August at Mirehouse Farm,[50] and late in October 1807 near Hope Pit, at the latter to a depth of 102 feet [=31 metres].[51] The owner of these mines was William Lowther, first Earl of Lonsdale (1757–1844). His agent was James Bateman (1749–1816) who wrote to Lord Lowther on 10 August 1807

> James Ryan the Irish Patent borer who wrote to your Lordship in London & and whose letter you sent me, we have employed near Myrehouse for the last three weeks in Boring, but he has only bored about 29 feet which at Mr Rawlings [then the best old style borers in the north[52]] price (being the price I agreed to pay him) comes to 29/-. His two men at 1 guinea each per week for three weeks will cost him 6 guineas and his own time and rods ought to make him as much [profit]. From the above statement your Lordship will see he has no chance to bore at the Rawlings prices. We have advanced him three Guineas on account, he wanted more but I told him he must earn it first. I expect he will leave us soon, as he must sink money very fast.[53]

This confirms that here Ryan's 'new technology' was simply seen as slower and more expensive. The fact that it could provide much better quality data for potential prospectors (as Farey had urged) was not thought relevant. In the context of a known mining area, like Whitehaven, we should perhaps not expect it to have been. But Bateman's letter ended 'we think him [Ryan] a great braggadochia.' This implies that Bateman also thought Ryan was full of empty promises and that he, and other members of the English mining fraternity, were prejudiced against such an Irishman, whom they feared had kissed the Blarney Stone too often.

The laissez-faire system of patronage then operative in Britain often confused the price of a scientific advance with its value. A particularly abortive trial for coal was then under way at Bexhill in Sussex, in hope of restoring the long faded iron industry there. This used old boring technology (supplied by the Birmingham engineer William Whitmore) under the authority of an Act of Parliament. But over several years it completely failed, despite an expenditure of £80,000 and a shaft many feet deep, simply because it was sited much too high up the stratigraphic series that Smith and Farey had worked out. It was later quoted as a classic example of the value of science in lowering the price of mining ventures.[54] Farey investigated this Bexhill attempt in person in the summer of 1806 and in August 1807 wrote of the specimens he collected east of Hastings which contained many detached pieces of bitumenised wood

> that were a [old style] augre-hole to be bored into it, and supplied with water, &c. something like the appearance of penetrating a coal vein, might be had in the borings; and it is this stratum, dipping under Bexhill, situate about 6½ miles to the westward, which in the opinion of Mr Farey has been there mistaken in the [old style] borings for a seam of coals, but which the improved

boring apparatus of Mr Ryan ... would have detected and saved, perhaps, a most unparallelled waste of money, in the measures now pursuing.[55]

In 1808 Ryan became involved with the Thames Archway Company, attempting to bore a first tunnel under the Thames at London, under the direction of Richard Trevithick (1771–1833). Trevithick's attempt was successful up to December 1807, when water and quicksand broke into the tunnel workings. In April 1808 a meeting of the directors reported the opinions of two eminent northern mining viewers William Stobart of Durham and John Buddle (1773–1843) of Newcastle-on-Tyne. This meeting then directed Ryan to make exploratory borings on the north shore.[56] The outcome of these borings is unknown as the tunnel was abandoned in 1809, due to the technical difficulties that continued to be met. But the choice of Ryan's apparatus shows that some were now choosing it for the good quality borehole data it gave, only then available by using Ryan's device, which was noted particularly for its ability to bore through 'gravelly soils'.

On 11 April 1808, Ryan demonstrated his apparatus to potentially its most important audience. This was a selection of 19 members of the Royal Society, including its President Sir Joseph Banks (1743–1820) and the Geological Society of London, including its President George B. Greenough (1778–1855). The demonstration was held near the house of the naturalist and artist James Sowerby (1757–1822) in Lambeth, near London. Here Ryan successfully bored 10 feet down with an eight-inch diameter corer. The list of those who witnessed the trial survives as does the detailed log of the strata penetrated.[57]

This meeting was also significant as Sowerby must then have introduced Ryan to the new methods of William Smith and John Farey who had been using fossils to identify particular strata against the Natural Order of rocks which they had previously elucidated. Smith had introduced himself to Sowerby the previous month when Sowerby became an enthusiastic convert to Smithian methods, unlike many of the gentlemen geologists of the Geological Society.[58] Smith's stratigraphic advances had already demonstrated the stupidity of some misguided attempts to find coal.[59] Ryan, probably guided by Sowerby, became an enthusiastic proponent of these new methods which brought some science to the 'art' of prospecting for stratified minerals. From June 1808 Ryan sent fossil specimens, letters and geological sections, including materials he had collected on his travels in Scotland, Ireland and at Coalbrookdale to James Sowerby,[60] then busy starting to compile his new work on the fossils of Britain.[61]

Ryan and Sowerby must also have entered into some form of business agreement as Ryan's later advertising broadsheets of about 1812 and 1817 note that orders for his services and the use of his Patent Boring Apparatus were to be placed at the 'British Mineral Collection, 2 Mead Place, St George's Fields, Lambeth', Sowerby's home address. Ryan had also by 1812 'formed a company of itinerant Miners, Drainers &c [to] execute orders in any part of the United Kingdom; make Estimates and Surveys of Estates, depicting their Mineral and Fossil contents on Maps or Models, contract for exploring Royalties by Boring; undertake to clear Coal Mines of inflammable or other Gas, however dangerous'.[62]

Ryan, probably less aware of the tensions then existing between the working class 'mineral surveyors', that he represented, and the gentlemen geologists of the Geological Society, also sent material to this society. His presents comprised some of the first stratigraphic collections that the society received. A particular collection of specimens in 'description of the Stratification extending from London to Stourbridge in Worcestershire' given, through the president, on 3 August 1808 was accompanied by a descriptive catalogue.[63] The latter survives to show both how well and how soon Ryan had espoused Smithian methods.

Ryan and Mine Ventilation

Edgeworth had noted in 1806 that there was a second potential use of Ryan's invention, to create passage ways to ventilate coal mines. This proved a highly political, rather than scientific, matter, in an England which was soon to be shamed into doing something about the massive mortalities in coalmines which yearly resulted from explosions. As a result of this situation, Ryan's device was soon 'highjacked' into the vitriolic debate which developed.

Ryan had first gone to Newcastle in May 1806 to try to get his device used there in coal prospecting, and John Buddle, the most eminent viewer there, even agreed to become an agent for it. But Ryan's attempts to also encourage its use for the better ventilation of coal mines, by using it to bore gas drains to rid the most affected mines there of mine damp, failed. Buddle refused to support the use of Ryan's new technology for ventilation in the North East,[64] retorting that there were very few deaths from such mine damp and that the existing ventilation system was adequate.[65]

In 1808 Ryan tried again under rather different circumstances in the South Staffordshire collieries. Here Ryan's method worked well. The thick coals of South Staffordshire allowed easier ventilation by his methods than the thinner and more faulted coal seams of the North-East which were worked over greater areas. Ryan was first employed here by Lord Ward and Dudley to clear his Netherton mine at Dudley. This brought Ryan to Dudley for the first time, which became his home for the rest of his peripatetic life. Ryan cleared this mine with two men and two boys within 20 days, between 3 and 27 December 1808. So successful was Ryan's system here that it was used throughout South Staffordshire and certificates were issued in 1811 recording Ryan's success in cutting gas-ways with his borer. One read

> we have no hesitation in saying that during 2½ years in the most dangerous works, without gas firing lines [previously used here to try and remove gas] or other expensive precaution Mr Ryan did not lose one life.[66]

Ryan's main supporters here were Lord William Ward (1750–1823) of Himley Hall, third Viscount Dudley & Ward and a major coalowner in the area and self-made Samuel Fereday (1758–1839) of Ettingshall Park, Sedgeley, a major iron and coal master and banker, a man both made by the Napoleonic War and broken by its concluding Peace in 1815.[67]

On 1810 Ryan lectured on his system at the Royal Institution in London, an organisation which was soon to become embroiled in the political debate on mine safety which involved scientists of the prestige of Sir Humphrey Davy.[68] An explosion in May 1812 at Felling colliery on Tyneside caused 92 deaths and brought a new urgency to the search for a solution. The Society for Preventing Accidents in Coal Mines was founded in Sunderland in October 1813, with Buddle still wholly opposed to the introduction of Ryan's methods.[69] Davy's invention of the safety lamp drew from the President of the Royal Society, Sir Joseph Banks, one of his most effusive letters.[70] The history of the safety lamp has involved many such strong opinions, and some myth making in the long war between science (Davy) and the practical man (George Stephenson).[71] The safety lamp was cheaper and thus seemed a more satisfactory solution than the improved mine ventilation advocated by Ryan. Ryan's own experiments showed that the Davy lamp could be dangerous and made to explode.[72] Posterity has duly confirmed how the introduction of the safety lamp generated a rise, not a decline, in the number of fatalities from gas explosions in English coal mines.[73]

A contemporary referred to the 'rejection of Mr Ryan's plans in the North' and the 'prejudices and mistaken interests which had coalesced in the North' in rejecting them. He noted that there were 'viewers who systematically oppose the introduction of invention' and how there 'are strong deeply-rooted prejudices here of more than fifty years' undisturbed growth, in favour of the existing system of ventilation, which the influence of any individual, be his genius what it may, can never remove'.[74] He specifically noted John Buddle, who opposed Ryan and supported the development of the Davy lamp.[75] Great prejudice continued in the North East against Ryan. The Rev. John Hodgson wrote in 1816 of having had to listen to 'three hours of Ryan's Irish eloquence' at a public lecture in Newcastle.[76] As late as 1828 Buddle noted to Hodgson that 'our friend Ryan is again amongst us ... The best way will be, probably to treat him with silent contempt, and allow him to exhaust his venom. The trade will not again be humbugged by him.'[77] The North East led British coal technology and such attitudes were taken as authoritative.

In 1816 the polarised politics of mine ventilation and safety erupted between the Royal Society and the Royal Institution (which favoured the Davy lamp) and the Society of Arts (which favoured Ryan). In 1816 the latter rewarded Ryan with their prestigious Gold Medal and the largest premium they had ever awarded, of 100 guineas.[78] Ryan published a *Letter* explaining his system in 1816[79] followed by an *Appeal* in 1817.[80]

In 1818 Ryan moved to Shropshire and leased mines at Middletown Hill in Montgomeryshire. He was elected a member of the Salopian Lodge of Freemasons based in Shrewsbury, as a 'Director of Mines', in hopes of new patronage.[81] But the mines caused him major financial embarrassment and he was still involved in litigation in 1840.[82] While at Middletown, Ryan erected at his own expense a school 'for instructing young Men in my system of Mining'. This was between 1819 and 1831 and must have been the first purpose-built mining school built in Britain.[83]

In 1824 Ryan published a new *Appeal*[84] and continued to urge the value of his ventilation system. In 1831 his first notice appeared in the *Mechanics Magazine*, whose editor noted that 'Mr Ryan and his system deserve more public consideration than they have hitherto received.'[85] On 6 July 1835 Ryan was called as a witness before the *Select Committee on Accidents in Mines* which the Government had now set up. This gave his history of his efforts with mine ventilation.[86]

The last years of Ryan's life were spent promoting his system of mine ventilation. The *Report of the South Shields Committee to investigate the Causes of Accidents in Coal Mines* of 1842 had been very favourable to his ideas and his work was now taken up by the *Mining Journal* which published a letter by another early pioneer in mine safety, Dr John Murray (?1786–1851), stating that

> we want ventilation, and a scientific system of working coal mines – not safety-lamps... In the *Report of the South Shields Committee appointed to investigate the Causes of Accidents in Coal Mines*,[87] we have an honest, honourable, and faithful tribute to the merits of Mr Ryan's plan. The award of the Gold Medal and 100 guineas, by the Society of Arts, in 1816, to Mr Ryan, was founded on the ample testimony, by practical men, of its triumphant success; and M. Boisse, the distinguished director of the Belgian mines, lauds the '*avantages incontestables*' of the system propounded by Mr Ryan. What further evidence do we require of the excellence of his plan?[88]

The Belgian referred to had been involved in a concourse held in 1840, to discover the best means of avoiding mine explosions.[89]

On 28 September 1844 a mine disaster at Haswell, near Durham took 95 new lives and inspired a new commission of inquiry under Charles Lyell and Michael Faraday. Ryan attended the inquest, coming all the way from Derbyshire, but was not allowed to speak by the coroner who recorded a verdict of 'accidental death'. Another disaster followed at Coxlodge Colliery in Northumberland in October 1844 after which Ryan's methods were again cited.[90] Ryan attended this inquest also, now at the expense of the Birmingham banker and philanthropist Benjamin Attwood (died 1874) who had been impressed by the value of Ryan's ideas. Once again Ryan was not allowed to speak at the inquest.[91] In a lecture at the Royal Institution in January 1845 Faraday paid tribute to Ryan '[whose] principle of withdrawing gas, I am glad to find, is not new among the coal-owners. Mr Ryan's method of ventilating mines is one which essentially depends upon drawing or draining of the gas from the mine. To my mind his principle seems very beautiful.'[92]

On 21 July 1847 Ryan died aged 77 at Dudley.[93] He was buried at St Thomas Church on 25 July.[94] He had latterly been receiving a yearly pension of £50 a year from the revenues of Lord Ward's mines.[95] All his papers and correspondence were left to the mining engineer Henry Johnson (1823–1885) of Dudley. Johnson was the founder in 1844 of the firm of Johnson, Poole and Bloomer, geotechnical engineers of Stourbridge and a few of Ryan's books and Henry Johnson's diaries still survive in their care. Johnson paid tribute to Ryan in 1862:

> happily with the introduction of the Davy many years ago, and the constant application of the late James Ryan, who was better known amongst the colliers by the significant cognomen of 'Count Sulphur', ventilation has vastly improved. His great theme was, 'remove the cause and the effect ceases', and by adopting his recommendation in this district of driving the air-heading on a level with the roof of the gaitroad, all the gas liberated is immediately carried off by the top air-heading. Previous to this recommendation, which there is not a shadow of a doubt emanated from him (as all his papers and correspondence, left with me at his death, clearly testify), the air-heading was driven universally in the benches, i.e. on the very floor of the mine.

Johnson concluded that Ryan was

> a departed genius, who, unhappily, in his day, was all heart and soul in endeavouring to alleviate the sufferings of the workmen, increasing the wealth of the coal owners, and yet, withal, did so little earthly good for himself or family, for his last moments were comparatively those of a pauper.[96]

A more recent historian in 1928 rightly noted he was a 'celebrated Irish mining engineer and one of the remarkable omissions from the *Dictionary of National Biography*'.[97] Ryan was clearly also a very brave man, as William Mathews noted in 1860:

> the present mode of ventilation is unquestionable the offspring of his genius The courage and perseverance with which Mr Ryan prosecuted his system were deserving of the highest praise, and the danger which he personally encountered and the resolution which he displayed were such that he was familiarly known amid the colliers by the name of 'hell-fire Jack'. He may be remembered by many coal owners of the present generation by the importunity with which he continually urged upon them the righteousness of coroner's juries finding a verdict of wilful murder in all cases of deaths from explosions in those collieries in which his system of ventilation was not acted upon.[98]

Ryan was the victim of polarised attitudes and geographies. He was popular among colliers but unpopular amongst mine owners, as his innovations involved expense. He was regarded as a genius in his adopted Black Country but as a liar, a charlatan and 'Irish' elsewhere. His truly original invention of the first boring apparatus which could recover cores from boreholes (now vital technology all over the world in search of minerals) was soon forgotten because, as Robert Bald (1776–1861) the Scottish mining engineer wrote in 1830:

> The impression which the public has at present regarding [his] mode of boring is, that it is very expensive . . . It is, however, but too well known, how much mankind are attached to old plans, strongly prejudiced against new ones, and that it is no easy matter to change particular habits; this may have operated against Mr Ryan's invention and have prevented success attending it.[99]

Such are the problems which have always faced the true innovator.

ACKNOWLEDGMENTS

The staff of the Cumberland Record Office, Carlisle, the Durham Record Office and the North East Institute of Mining and Mechanical Engineers in Newcastle provided all possible assistance. Ruth Heard (Dublin) gave much appreciated help in getting me both details and access to GCI records. I thank Gordon Herries-Davies (Nenagh), Valerie Ingram (Dublin), Colin Knipe (Brierley Hill), Nick Lee (Bristol), Linde Lunney (Dublin), Colain Macarthur (Letterkenny), John Thackray (London) and Patrick Wyse-Jackson (Dublin) for their kind help.

NOTES

1. R.V. & P.J. Wallis, *Index of British Mathematicians Part III* (Newcastle-upon-Tyne, 1993), p. 119.
2. R.L. Galloway, *Annals of Coal Mining* (London, 1898), p. 411.
3. In a letter to Josiah Wedgwood junior dated 6 September 1831, Wedgwood archives 21880–29, Keele University Library.
4. Ryan's printed flyer 'To the Coal-Owners of the Tyne and Wear' dated 15 November 1843, Bell collection, North East Institute of Mining and Mechanical Engineers, Newcastle.
5. Memorandum on mining, Wedgwood Archives 21874–29, accompanying his letter to Josiah Wedgwood junior dated 3 August 1831, 21878–29, Keele University Library.
6. *Transactions of the Irish Academy*, 2 (1789): pp. 157–170.
7. Archives of Johnson, Poole & Bloomer, Brierley Hill.
8. *Report from the Select Committee on Accidents in Mines* (London, 1835), p. 204 (answer 2830).
9. Ruth Heard, *The Grand Canal of Ireland* (Newton Abbot, 1973), pp. 143–7.
10. W.Tighe, *Statistical Observations relative to the County of Kilkenny* (Dublin, 1802), plate IX.
11. G. Bowie, 'Early Stationary Steam Engines in Ireland', *Industrial Archaeology Review*, 2 (1978): pp. 168–74, p. 168.
12. GCI Minute Book, volume 28, entry for 24 April 1804. The earlier volume (27) does not survive.
13. Heard, *The Grand Canal of Ireland*, p. 144.
14. Obituary in *Proceedings of the Institution of Civil Engineers*, 3 (1844): pp. 14–5.
15. GCI Minute Book, volume 30, entry for 27 December 1804.
16. *Report from the Select Committee on Accidents in Mines*, p. 204 (answers 2827–9).
17. Ryan, 'To the Coal-Owners of the Tyne and Wear'.
18. *Proceedings of the Dublin Society*, 40 (1803–4): p. 83.

19. *Proceedings of the Dublin Society*, 41 (1803–4): p. 15.
20. Ryan, 'To the coal-Owners of the Tyne and Wear'.
21. Number 2822.
22. C.J. Jackson, *A History of the Development of Drill Bits and Drilling Techniques in Nineteenth Century Britain* (Open University Ph.D., 1983), pp. 215–8.
23. *Monthly Magazine*, 19 (1805): pp. 368–9.
24. *The Retrospect of Philosophical, Mechanical, Chemical . . . Discoveries*, 1 (1806): p. 85.
25. *Magazin aller neuen Erfindungen*, 6 (1806): pp. 223–7 and plate 4.
26. Ryan, 'To the Coal-Owners of the Tyne and Wear'.
27. William Nicholson's *Journal of Natural Philosophy etc*, 15 (1806): p. 85.
28. *The Retrospect of Philosophical, Mechanical, Chemical . . . Discoveries*, 2 (1806): p. 410.
29. Copy in Bell papers, volume 6, p. 162, British Geological Survey archives, Keyworth.
30. B. Burke, *The Landed Gentry* (London, 1871), pp. 1081–2.
31. B. Burke, *The Landed Gentry* (London, 1871), p. 1164.
32. *Dublin Gazette*, 8 May 1802.
33. A. Rees, 'Coal', in *The Cyclopaedia*, (London, 1807).
34. *GCI Minutes*, volume 31, entry for 15 April 1805.
35. *GCI Minutes*, volume 31, entry for 4 May 1805 (pp. 229–45).
36. Patent numbers 2469 (1801) and 2607 (1802).
37. James Millar, *The Natural History of the Mineral Kingdom by John Williams* (Edinburgh, 1810), volume 2, p. 356.
38. Michael Flinn, *The History of the British Coal Industry*, II (Oxford, 1984), p. 72.
39. H.S. Torrens 'The History of Coal Prospecting in Britain 1650–1900' in *Energie in der Geschichte* (Dusseldorf, 1984) pp. 88–95 and 'The History of Coal Prospecting in Britain – a neglected subject', *Geology Today*, 2, no. 2 (1986), 57–8.
40. Francis G. Dimes, 'Correspondence', *Geology Today*, 2 no. 5 (1986), 138.
41. GCI Minutes, volume 31, entries for 15 April to 9 May 1805.
42. Tyne and Wear Archives, 1589/102–6.
43. Anne Plumptre, *Narrative of a Residence in Ireland . . . in 1814 and 1815* (London, 1817), p. 98.
44. Board of Agriculture, Register of letters received, B XII, and Minute Book fair copy, B VII, University of Reading, Institute of Agricultural History. See also Ryan, 'To the coal-Owners of the Tyne and Wear'. The Board's testimonial is reprinted in Ryan's 1812 testimonial cited above, note 29.
45. Rees, 'Coal'.
46. H.S. Torrens & T.D. Ford, Introduction to the reprint of John Farey's 1811 *General View of the Agriculture and Minerals of Derbyshire* (Matlock 1989).
47. Printed Broadsheet of *c.* 1817, of 'J. Ryan, F.A.S. Mineralogical Surveyor, Director of Mines &c' in the author's collection.
48. MSS in Dept of Geology, Trinity College Dublin.
49. Testimonial cited above, note 29.
50. Cumberland Record Office, D Lons. W, Whitehaven district, Boring journal 1776–1842.
51. Cumberland Record Office, D Lons. W, 'Journal of Incidental Borings not connecting with the Whitehaven Coal or Lime Field', no. 39.
52. Flynn, *History of the British Coal Industry*, p. 72.
53. Cumberland Record Office, letter of Bateman to Lowther, 10 August 1807.
54. John F.W. Herschel, *A Preliminary Discourse on the Study of Natural Philosophy* (London, 1831), p. 45.
55. A. Rees, 'Colliery', in *The Cyclopaedia*, (London 1807).
56. Francis Trevithick, *Life of Richard Trevithick* (London, 1872), volume 1, p. 269.

57. James Sowerby archive, Natural History Museum, London.
58. L.R. Cox, 'New Light on William Smith and his Work', *Proceedings of the Yorkshire Geological Society*, 25 (1942): pp. 1–99, pp. 53–4.
59. H.S. Torrens, 'William Smith's "New Art of Mineral Surveying" and the "Brewham Intended Colliery": a failed attempt to find coal in Southern England between 1803 and 1810' in G. Gohau, ed., *Livre Jubilaire Francois Ellenberger* (Paris, 1996), in press.
60. Four of James Ryan's letters to Sowerby, dated 1808, 1809, 1830 and undated, are in the Eyles archive, Bristol University Library.
61. See James Sowerby, *The Mineral Conchology* (London, 1812–1846), 7 vols.
62. Testimonial cited above, note 29, and the broadsheet, note 47.
63. Geological Society archives MUS 1/94 and 'Donations to the Cabinet of Minerals', *Transactions of the Geological Society of London*, 1 (1811): p. 409.
64. *Tracts on the Necessity of Legislative Interference in protecting the Lives and Health of the Miners against Colliery Explosions and the Injurious Effects of Badly Ventilated Mines* no. 2 (8 December 1849), p. 13 and ref. 38, p. 135.
65. *Report from the Select Committee on Accidents in Mines*, answer 2834.
66. Testimonial cited above, note 29.
67. Fereday fled to France in 1821 where he died an uncertificated bankrupt in 1839 (*Staffordshire Advertiser*, 6 April 1839, p. 3).
68. M. Berman, *Social Change and Scientific Organization: The Royal Institution 1799–1844* (London, 1978), pp. 175–182.
69. J.H.H. Holmes, *A Treatise on the Coal Mines* (London, 1816), pp. 146–155.
70. T.G. Vallance, 'Jupiter Botanicus in the Bush', *Proceedings of the Linnean Society of New South Wales*, 112 (1990): pp. 49–86, pp. 53–4.
71. A.R. Griffin, 'Sir Humphrey Davy: his Life and Work', *Industrial Archaeology Review*, 4 (1980): pp. 202–13, p. 202.
72. J.H.H. Holmes, 'On Safety lamps for Coal-Mines', *Annals of Philosophy*, 8 (1816): pp. 129–131.
73. David Albury and Joseph Schwartz, *Partial Progress* (London, 1982) and A.F.C. Wallace, *The Social Context of Innovation* (Princeton, 1982), p. 115.
74. Holmes, *A Treatise on Coal Mines*, pp. 88, 140 & 153.
75. T.Y. Hall, *On the Safety Lamp for the Use of Coal Mines* (London, 1854), p. 14.
76. Durham Record Office, NCB I/JB/1496, letter to Buddle, 17 June 1816.
77. J. Raine, *A Memoir of the Rev. John Hodgson* (London, 1858) I, pp. 170–1.
78. 'The Gold Medal and One hundred Guineas . . . to Mr James Ryan . . . ', *Transactions of the Royal Society of Arts*, 34, (1816): pp. 94–121.
79. *A Letter from Mr James Ryan . . . on his Method* (London, 1816).
80. *The Appeal of James Ryan . . . to Proprietors of Collieries . . .* (Birmingham, 1817).
81. A. Graham, *A History of Freemasonry in The Province of Shropshire* (Shrewsbury, 1892), p. 232.
82. *Mining Journal*, 10 (no. 245), 2 May 1840, p. 138.
83. Keele University Library, 21874–29.
84. *The Appeal of James Ryan . . . to Proprietors of Collieries* (Stoke-upon-Trent, 1824).
85. *Mechanics Magazine*, 15 (1831), p. 461.
86. *Report from the Select Committee on Accidents in Mines*, answers 2824 to 2940.
87. Reprinted by James Mather, *The Coal Mines, their Dangers and Means of Safety* (London, 1858), pp. 23–72.
88. *Mining Journal*, 13 (23 December 1843), p. 419.
89. A.-A.-M. Boisse, 'Memoire sur les Explosions dans les Mines de Houille' pp. 35–140 of *Des Moyens de Soustraire l'Exploitation des Mines de Houille aux Chances d'Explosion* (Bruxelles, 1840), pp. 62–8, 130.
90. *Times*, 17 October 1844, p. 6e.
91. *Times*, 1 November 1844, p. 6d.

92. *Civil Engineer and Architects Journal*, 8 (April 1845), p. 118.
93. *Mining Journal*, 17 (July 1847), p. 346.
94. Parish Register, Dudley Library.
95. Dudley Archives Office, Dudley Estate Archive, V/2/4.
96. H. Johnson, 'On the Mode of Working the Thick or Ten Yard Coal of South Staffordshire', *Transactions of the North-East Institution of Mining Engineers*, 10 (1862): pp. 183–96, p. 194.
97. A.J. Hawkes, *Jubilee Exhibition of Early Mining Literature* (Wigan, 1928), p. 34.
98. 'On the Ten Yard Coal of South Staffordshire and the Mode of Working', *Proceedings of the Institution of Mechanical Engineers*, (1860): pp. 91–120, p. 105.
99. Robert Bald, 'Mine' in D. Brewster, ed., *The Edinburgh Encyclopaedia*, 14 (1830): pp. 314–378, pp. 330–1.

Chapter 6

Technical Education and the Application of Technology in Ireland 1800–1950

W.G. Scaife

INTRODUCTION

As the eighteenth century drew to a close it was possible to observe the acceleration which had taken place in the application of technology in many European countries. This was most obvious in England where the mastery of steam power had given industry a great advantage over continental rivals. In Ireland too a period of relative peace had brought a great increase in prosperity, aided in part by new trading possibilities. Many large scale infrastructural projects had been undertaken. The Newry canal for example, which was opened in 1742 for the transportation of coal, was the first such venture in the British Isles.[1] Other large projects were undertaken which called for an input from trained engineers, such as harbour construction, bridge and road building, and after 1835, railway construction. As the nineteenth century progressed manufacturing industries increasingly offered opportunities to suitably educated engineers capable of applying new technology.

For many trades and professions the time honoured way of learning was by way of apprenticeship to an experienced practitioner, whether it be a metal worker or a surgeon. Gradually the value and importance of providing a suitable educational foundation was recognised. In this regard developments in the British Isles diverged significantly from the practice which became established in other European countries. Because of its history and proximity to England, Ireland resembled England in most respects. But because it was ruled as a colony, with an all powerful administration in Dublin answerable only to the parliament in Westminster between 1801 and 1922, there were aspects of the educational system which emerged which had something of a continental style about them.

The extent to which the provision of suitable technical education could benefit industry was dependent on attitudes of society. To take an extreme example, consider one of the very first attempts to mount a scientifically based engineering course which was made at the newly created university of Durham in 1838.[2] Within thirteen years it had to be abandoned because of a lack of demand for its graduates. What makes this all the more remarkable is the proximity of Durham to the well springs of the industrial revolution on Tyneside, with its coal mines, shipyards and railway works. It is clear that the local employers did not value university educated graduates. In this respect the negative influence of social attitudes in a section of English society can be discerned.

DIFFERENT CULTURAL TRADITIONS

The quickening pace of technological progress brought with it a rapid increase of population. This was felt in Ireland as much as in England. By 1841 the population exceeded eight million, which was close to one half that of England and Wales at the time, and comparable to that of a great state like Spain. Before this date reliable data are difficult to establish but growth was unusually rapid. As Cullen[3] comments 'Ireland . . . [experienced] a sustained demographic expansion which, if the period 1600 to 1845 is taken as a whole, has few parallels in pre-industrial Europe'. The sharp increase in the pace of development was associated with rapid progress of scientific studies. A classic example of this was the successful construction of the first steam engine by Newcomen in 1712. This put to practical use the scientific knowledge gained in studies carried out by men like Torricelli and Robert Boyle, a son of the Earl of Cork.

The industrial revolution in England was the work of ingenious and energetic craftsmen, many of whom like Newcomen were nonconformists. As such they were often literate. They valued education because it gave them direct access to the Bible. They were independently minded in that they refused to conform to the established church. This nonconformism incidentally denied them access to university education until the middle of the nineteenth century. That independence of mind which was so valuable among artisans was also exhibited by the capitalists whose wealth was essential for the exploitation of technological developments. These wealthy land owners had already demonstrated their ability to limit the power of the Stuart monarchy, with a rebelliousness which culminated in the Glorious Revolution in 1688. Henceforth attempts by the central government to take initiatives of the kind possible in other European states which in general lived under more authoritarian regimes, met strong resistance. Many had grown wealthy through sharing in the opportunities for exploitation afforded by an age of imperialism which stretched back to Elizabethan times and which was to reach its apogee at the end of the nineteenth century. Their interests were in commerce, and their attitude to industry was to keep involvement at arms length. The social behaviour which these attitudes produced was to exert a strong influence on the development of technical education. Inevitably Ireland was powerfully affected by this too, not merely because of its proximity to England, but by a significant influx of Englishmen.

On the continent of Europe it was France which pioneered many developments in technical education.[4] In 1747 the École des Ponts et Chaussées had been founded to educate civil and military engineers for the service of the state, and the state took an active interest in providing education.[5] The cataclysm of the French Revolution cleared the ground for radical measures in this area. It was around this time that the profession of scientist emerged and the help of scientists was sought as part of the war effort. Central to this was the newly reformed Académie des Sciences.[6] In 1794 the École Polytechnique was established to provide a preliminary scientific and mathematical education for the *écoles d'application* such as the École des Ponts et Chaussées and the École des Mines. It was to be a model for others to emulate in other countries later in the century. The names of those who taught or were graduates of the institute form a glittering collection of pioneers in science and mathematics. To be successful such institutions required the existence of suitable second level schools which could prepare candidates for admission. This demand was met to some degree by the *lycées* created under Napoleon.

A different kind of institution created before the Revolution, the Conservatoire des Arts et Métiers, was founded as a museum to house models, tools, designs and books, which were a gift from a wealthy individual. By 1819 courses of lectures were being offered there in mechanics, chemistry applied to the arts and in industrial economy. Evening lectures were given to facilitate workers who wished to attend.[7] Again it was to prove an inspiration to others in later years. In 1829 a group of academics were funded by a wealthy entrepreneur who established the École Centrale des Arts et Métiers. It aimed specifically to meet the educational needs of industry. Other lower level institutions had been created by Napoleon to educate industrial foremen and managers. The influence of such innovations was spread to Spain,[8] Belgium[9] and Italy[10] during the period of French occupation. England on the other hand, was shielded by virtue of her military success.

During the nineteenth century the engineering profession was divided between those who were needed to meet the demands of state service, largely civil engineers in the modern meaning of the term, and those who would work in manufacturing industries. While the former group was well catered for, the latter group had much longer to wait, and indeed they made little immediate impact at first on industry, which was still at what may be described as the first phase of the Industrial Revolution.[11] The importance of the École Polytechnique was that it provided a model for the teaching of science and mathematics to aspiring engineers and other public servants, and ensured a public acceptance of the educational importance of these subjects. English engineers, although denied such systematic education in mathematics and science, had made contributions to technological progress which were appreciated by many in France. For example C. Dupin, himself a graduate of the École Polytechnique and one of the professors at the Conservatoire des Arts et Métiers, wrote warmly of it.[12] The fact is that while the engineer's design must be scientifically based, it must also be concretely realised, and so his education must also involve practical experience. Both features must be provided for, but the balance between them has varied in different countries and at different times.

State institutions like the Royal Military Academy at Woolwich in England never achieved the status of their French counterparts, and the older universities of Oxford and Cambridge showed no interest in engineering or technology until the end of the nineteenth century.[13] The first steps to teach engineering in a university were taken by the Anglican King's College in the new University of London, when a course in civil and mechanical engineering was proposed in 1838. At the same time University College in the University of London, which had been established to cater for Dissenters, appointed the Irish railway engineer C.B. Vignoles to a chair of civil engineering. The abortive venture in Durham has been mentioned already. Nowhere was there the equivalent of the concentration of scientists and mathematicians in institutions such as the École Polytechnique, nor was there a body of graduates who held scientific studies in high esteem.

In England efforts to provide elementary schooling during the first part of the nineteenth century had totally failed to meet the needs of the rapidly growing population. The evidence given to the Parliamentary Select Committee on Scientific Instruction[14] in 1867 revealed very great deficiencies in the provision of education for children in England, especially when compared with the practice in other European countries such as Prussia. It was organised on a local basis, and religious tensions between the Established Church and Dissenters delayed the introduction of state funding until after the passage of the Education Act in 1870. Secondary schools tended to promote a strong classical bias with little or no scientific teaching, and it was not until the 1902 act of parliament that a public secondary school system was set up.[15] Among those involved in manufacturing, such as artisans and managers, apprenticeship was the usual route for progress, attendance at classes, if any, being normally after a long working day. From 1820 a voluntary movement developed in response to demands from this section of the working population for scientific education. It led to the establishment of Mechanics Institutes in many locations.

By the early part of the nineteenth century the profession of civil engineer had become well established, the Institution of Civil Engineers was founded in 1818. As one of the founders, H.R. Palmer said:

> An engineer is a mediator between the philosopher and the working mechanic, and, like an interpreter between two foreigners, must understand the language of both . . . Hence the absolute necessity of his possessing both practical and theoretical knowledge.[16]

Admission to the profession was by way of apprenticeship, with a system of fees which could be substantial.[17] But those involved in the use and manufacture of machinery, mechanical engineers, or 'mechanics', struggled to have their concerns recognised, and were induced to set up the Institution of Mechanical Engineers in 1847.

SCHOOLING IN IRELAND

The decision for the state to take control of primary education in Ireland was made much earlier than in England. In part the introduction of a state funded National School system in 1831 may well have been motivated by a wish to limit the spread of disaffection which was feared from the existing system of 'hedge schools'. Nevertheless it certainly did hasten the spread of literacy.[18] Its implementation had more of the character of the action of a continental state than of a normal response of the London parliament. The establishment of Mechanics Institutes in England was copied in Ireland.[19] By 1825 they were to be found in Armagh, Belfast, Cork, Galway, Limerick and Waterford. Eventually in 1854 the Department of Science and Art was created in London. It provided funding to the teachers of students who passed the examinations which were set by the department. The benefits of the system of National Schools which had been introduced in Ireland was such that it was Irish candidates who took a major share of the prizes awarded on the results of these examinations.[20] During the latter part of the nineteenth century the Catholic community put enormous effort into providing secondary education.[21] In large measure this was free of state direction. Second-level schools of all denominations in Ireland, as in England had a strong classical bias with little teaching of science. Even as late as 1931 only 14% of candidates for matriculation at the National University took chemistry and 4% physics.[22]

The Act of Union in 1801 secured the dominance of the established church in Ireland. Nevertheless, as in England there were dissenting religious groups which cherished their own approach to education and displayed their own outlook on industrial development. Quaker families were to be found scattered throughout the island, while Presbyterians, with strong connections with Scotland were concentrated in the north east. The largest group, Roman Catholics were also excluded from a full participation in public life. Despite the public funding of the seminary at Maynooth in 1795, European influences remained strong among Roman Catholics.

EARLY INDUSTRIES IN IRELAND AND INTEREST IN SCIENCE

The eighteenth century saw considerable small scale industrialisation spread throughout Ireland, in the form of iron works, textile mills, breweries and the like.[23] Water power was extensively harnessed, while the steam engine was quickly adopted for mining and other purposes. It must have appeared to many that provided coal, iron and minerals could be exploited locally, it would be possible to emulate England and provide industrial employment for the surging population. An interest in the potential to benefit from the application of science to agriculture and industry had led to the creation of institutions with government funding, like the Royal Dublin Society in 1731, and the Royal Cork Institution in 1799.[24] This is not altogether surprising since one aim of the earlier policy of plantation had been to encourage the immigrants to introduce new industries and to improve methods of farming.

In the north, Presbyterians had successfully established the Belfast Academical Institute in 1814 to form a secondary school and to provide university level courses, which included teaching in science for clergymen and laymen alike.[25] Thomas Andrews who was appointed to the chair of chemistry, was in time to achieve eminence.[26] Similarly Nicholas Callan who was appointed to the chair of natural philosophy in the Catholic college of Maynooth in 1826 achieved fame as an experimental physicist.[27] In Trinity College at this time, many wide ranging reforms were introduced under the leadership of Provost Bartholomew Lloyd. As a mathematician himself, he warmly espoused the work of the French mathematicians, and also introduced general reforms which were to bear abundant fruit later in the century.[28] An interest in the potential for mining to create wealth and employment had led to the establishment in 1831 of the Geological Society. Sir Robert Kane FRS, one of the first Roman Catholics to graduate from Trinity College, had published his *Industrial Resources of Ireland*[29] in 1844, and was appointed the first director of the Museum of Economic Geology in 1845. Like many others he was aware of developments in France and with Robert Mallet and other businessmen sought unsuccessfully, to set up an institution on the lines of the French École Centrale des Arts et Manufactures, with their own funding.[30] Up to this point the engineering profession in Ireland had followed the pattern in England, with apprenticeship the normal route for training. In 1835 the Civil Engineer's Society of Ireland was formed by Colonel John Fox Burgoyne to serve their needs.[31] Inspection of membership applications shows that men with a formal engineering education remained very much a minority of its membership during the nineteenth century.

In Ireland the funding of the Presbyterian and Catholic colleges in Belfast and Maynooth did not satisfy the demand for establishments independent of church affiliations. In 1835 Thomas Wyse, a Trinity graduate and one of the first Catholic members of the Westminster parliament to be

Technical Education and the Application of Technology　　　　　　　　　　91

elected after emancipation, was appointed chairman of a select committee of inquiry into education.[32] Wyse and Robert Kane, like Lyon Playfair in England could see the clear need for action if industrial progress was to be sustained. Wyse's report published in 1838 was very comprehensive with proposals for primary and secondary schools, agricultural and professional schools, and contained a suggestion for four provincial colleges. It did not find favour although it did prove to be a catalyst for the creation in 1845 of three Queen's colleges. located in Cork, Galway and Belfast, they were united in the Queen's University of Ireland in 1850. Having made up its mind, Peel's government could implement its decisions much more decisively than was possible in England. Once again these actions of the London parliament were reminiscent of the actions of a continental state.

INITIATIVES BY UNIVERSITIES IN IRELAND

In its internal affairs Trinity College was like Oxford and Cambridge, largely independent of external pressure, and change came from within. In 1813 Bartholomew Lloyd was appointed professor of mathematics, being elected Provost in 1831. He introduced the work of mathematicians like Poisson and Laplace to the curriculum.[33] His son, the Rev. Humphrey Lloyd was appointed professor of natural and experimental philosophy. In April 1841 the latter, together with James McCullagh, professor of mathematics and Thomas Luby, lecturer in mathematics, put a proposition to the board of the college. It was for the establishment of a School of Civil Engineering and Architecture. By November of that year Lloyd was giving the inaugural lecture to the first students. Robert Kane's efforts and the Wyse committee proposals which included plans for a Dublin polytechnic,[34] may have helped secure such speedy consent from the board. It was nevertheless a remarkably swift move for an institution already 250 years old. In that it was a move that was not dictated by the state, it shared a resemblance to what was happening in the new universities in England rather than in Europe.

Before the School of Engineering was opened at Trinity College, the Board sent Humphrey Lloyd to Paris as he wrote 'to acquaint myself with the French plan of instruction for engineers, with a view to borrowing whatever will bear transportation to a British climate for our Dublin school, for the success of which I am chiefly responsible'.[35] As his inaugural lecture revealed, he had a good grasp of the task. To quote: 'where the practical applications of science are concerned, the sciences themselves must be systematically taught. It is in the universities, the established schools of science, that such applications may best be unfolded.'[36] His opinion was that whereas in England 'practice has been insisted on almost to the exclusion of theoretical knowledge ... the Polytechnic School of France ... is decidedly too abstract in its studies.'[37] One feature of the French system had his approval, the *projet*: 'an engine is to be constructed for some special work, a suspension bridge is to be thrown across a river, or a manufactory of some kind is to be constructed under given conditions. He (the student) is required to put himself in the place of the professional engineer, to whom such an enquiry is addressed by a capitalist.'[38] He had a good grasp of the central role played by steam power in changing contemporary life, and was able to quote from the latest performance records of English pumping engines. Much of his address was devoted to justifying the presence of such a practical subject in a university. His view of engineering obviously was inclusive and not limited to what is now called civil engineering. From the start the school was well funded, with grants for models and housing for collections of minerals. The construction of an impressive building was authorised, being completed in 1857.[39] It was to be late in the century before other engineering schools in the British Isles could compete on an equal footing. The adoption of engineering education within the university was no doubt some advance, but it could not remove the negative social reactions of members of the wealthy establishment towards industry. Very soon this resulted in a concentration on civil engineering in the narrow sense.[40]

The consequence of admitting engineering students to Dublin University was that all should complete the first year of the arts degree. This involved mathematics, logic and classics, and this of course restricted entry to those with appropriate schooling. Two years of study, increased to three in 1845, which included work in the drawing office and practical exercises, led to the award of a Diploma in Civil Engineering. Although degrees in engineering were not introduced until later (1860, for the master's, M.A.I. degree, and 1872, for the bachelor's B.A.I.), the majority of engineering students also completed the B.A. degree course.[41] The training in communication and logical reasoning which were components of the arts course did constitute a valuable asset for men who would often have to rely heavily on empirical knowledge. In later years the requirement of the B.A. degree would prove to be valuable, for it ensured that in Ireland graduate engineers were seen to be on a par with other professionals like lawyers and clergymen who were in the public eye. This status was also enjoyed by the graduates of the other university colleges.

From the start, all the Queen's colleges offered courses for a diploma in Civil Engineering.[42] The matriculation requirements to some extent paralleled the arts requirement of Trinity, but did not require Latin or Greek, and the first year course included a modern language. In other respects the courses in the two universities were quite similar. In 1881 the Queen's University of Ireland was replaced by the Royal University,[43] which was an examining body for the students of the Queen's Colleges. The Catholic University which had been established on a voluntary basis initially to teach medicine, was now renamed University College Dublin and received public funding through the Royal University. Engineering was

Technical Education and the Application of Technology

not taught there until after 1909. The Royal College of Science had been formed from the Museum for Irish Industry in 1865.[44] It had evolved in a somewhat similar way to the College of Science at South Kensington being planned to be a supplier of science teachers.[45] It offered courses in engineering, but came directly under the control of the Department of Science and Arts.

The position in Prussia was different. There, candidates for state posts had to be graduates of the Bau-Akademie which required a classical school education, and enjoyed a similar social status to university graduates. But mechanical engineering was taught in institutions which required entrants to have mathematics and science though not Latin and Greek.[46] These graduates had to struggle for esteem in the face of a system favouring the humanities.[47] In England the engineer's position was far worse, and was not clearly distinguished in the public mind from that of craftsmen. This difference between England and Ireland in public perception has remained even to the present day. In Ireland, in contrast to England, the engineering profession has retained a strong attraction for school leavers.

Employment of graduate engineers in Ireland

Where did the engineering diplomates and degree holders find employment? A study conducted in 1853 showed that more than half of the sixty or so graduates of Trinity College had emigrated, mostly to Canada, India and Australia.[48] Only three had found employment in England, which is understandable in terms of the fate of the course in Durham, and the same number went to Europe. The occupation of just under a half was given as civil engineer/surveyor. An almost equal number were listed as railway engineers. None were engaged in manufacturing. Many were to be appointed to chairs in a number of institutions in Ireland and abroad. This general picture changed little in succeeding years. When the Parliamentary Commissioners visited TCD in 1851, Professor Apjohn volunteered that 'it would benefit the School and society at large if preference were given to diploma holders when selecting engineers for public service'.[49] Despite this it required a major effort on the part of the College Board in 1859 to persuade the Indian Civil Service to accept the diploma as equivalent to an apprenticeship. They were unsuccessful in gaining access for the diplomates to the Royal Military Academy at Woolwich.[50]

By modern standards the annual output of engineering graduates and diplomates was small, but until the end of the century, the Trinity engineering school dominated numerically the output of diplomates and graduates.[51] While graduate engineers failed to find employment in manufacturing and the shipbuilding industry, they did fill a valuable function in various public services. The Office of Public Works, the Dublin Port Authority, local authorities and the many railway companies all offered openings for civil engineers, and the profession had a good public standing.

INDUSTRIAL DEVELOPMENT IN IRELAND

During the nineteenth century industrialisation spread, most obviously in the north east around Belfast but also throughout Ireland. The strong social ties with Scotland helped the transfer of know how from the industrial centre on Clydeside.[52] By the middle of the century the output of Belfast had left behind the shipyards in Cork and other ports.[53] Its textile industry, flourished and with it the manufacture of machines to supply it. Naturally employers in Ireland tended to share the same attitudes as their counterparts in Britain and so virtually no graduate engineers were involved in the great industrial achievements in the north. It should be said that a not dissimilar situation was experienced in other European countries where university educated engineers were slow to find employment in industry.[54] It has been pointed out that the situation changed as the Industrial Revolution entered a new phase in the later part of the nineteenth century, with the emergence of industries which depended more immediately on a scientific input. Examples would include the manufacture of inorganic chemicals such as dyes, and electromechanical machinery such as turbo generators. In countries like Germany and Switzerland the structured organisation of primary and secondary schools facilitated the emergence towards the end of the century of technical institutes with a close relationship with industry.[55]

Some rather unique industrial developments were associated with a number of Quaker families.[56] These ventures were often remote from large population centres, and employed large work forces in well-built factories, with provision of model schooling and housing for the work force. The Malcolmsons at Portlaw in County Waterford employed 1,800 hands in a cotton mill. Considerable engineering skills were displayed by the Goodbodys at their jute mills in Clara. At Bessbrook the Richardsons established a linen mill in 1845 which at its peak employed 3,500 hands. In 1883 a pioneering scheme harnessed hydraulic power to generate electricity which was used to run an electric railway from the mills in Bessbrook to Newry. Many Quakers, although lacking university education, proved to be able engineers; the Grubb family, for instance, manufactured telescopes and optical instruments for sale world-wide. In general the Quakers' projects were commercially successful. One reason relates to their religious beliefs. Because they refused to swear an oath, careers such as the law, the army and of course the church were closed to them. Therefore many able and educated individuals turned their hands to industrial enterprises, who in families belonging to the establishment would have found more socially acceptable outlets. This episode lasted for a century and served to highlight the loss to industry when able and educated people are deterred from participating. It also pointed up the commercial benefits of having well-educated personnel in management positions.

PROVIDING FOR THE TEACHING OF SCIENCE IN SCHOOLS

As the Industrial Revolution moved into a second phase, the need for systematic scientific education of those involved in industry, whether as workers or as owners or managers, was becoming clear to some industrialists in England. Engineering courses were being offered in a dozen or so provincial colleges by the last quarter of the century.[57] In 1878 in London, as a result of local initiative, the City and Guilds Institute was established.[58] It opened technical colleges, and more importantly, provided course outlines and ran corresponding examinations for technicians throughout the whole of the United Kingdom. In Belfast and in Dublin strenuous local efforts were made to establish technical schools.[59] Dublin Corporation created the Kevin Street school in 1887. However it was not until the Agriculture and Technical Education (Ireland) Act was passed in 1899, after a parliamentary campaign led by Horace Plunkett, that the State took a hand and funding was made available. Technical schools were then quickly established throughout Ireland, and for the first time involved local authorities. They provided a range of courses for ambitious and able apprentices, though these were for the most part taken in the evenings after a day's work. At that time in Ireland as in England, it was usual for children leaving elementary schools to be apprenticed and set to learn their trade by observation of skilled workers. The luckier ones could attend night classes. Contemporary perceptions of the place of technical education in the scheme of things are evident in the poster prepared for the Belfast Municipal Technical Institute in 1907.[60] Privileged apprentices in the shipyard for example could attend evening or even day-release classes. This was the preferred educational route for senior personnel. It is clear that formal scientific education of men destined for industry was very restricted. In a discussion 'On the training of young engineers' held in 1900 on Tyneside, this situation was discussed and unfavourably compared with the State run systems in France, Germany and Japan.[61]

DEVELOPMENTS IN ENGINEERING COURSES

The Royal University gave way in 1908 to the National University and the Queen's University of Belfast.[62] The chief development around this time was the emergence of electrical and mechanical engineering courses. These were offered through co-operation with the Royal College of Science in Dublin, which had recently been provided with a fine new building. Attempts in Trinity College, led by the eminent physicist Prof. G.F. Fitzgerald, to promote the teaching of electrical engineering were not successful. Although laboratories were built and equipped to teach electrical and mechanical engineering, Prof. Alexander was not interested in developing the School of Engineering in that direction.[63] As in the case of the original proposals to establish the School of Engineering in 1841, it was among the scientists and mathematicians that initiative and clear sightedness were to be found. Eventually provision was made for students of both Trinity and of University College to take such courses in the Royal College of Science. The latter ceased to exist as an institution outside the university system when it was absorbed by merging with University College Dublin in 1926. This ended for many years the experiment of teaching engineering outside the university system in Ireland. In Belfast too arrangements were made in 1912 for engineering students from Queen's University Belfast to have access to courses in mechanical and electrical engineering at the Municipal Technical Institute. These steps were contemporary with the creation of the Imperial College of Science and Technology in London,[64] and the flourishing of the Technische Hochschulen in Germany.[65] The Royal College of Science in Dublin and the Belfast Municipal College of Technology failed to develop that close relationship with manufacturing industries which characterised the German Technische Hochschulen and failed to become involved in postgraduate teaching. This was no doubt because industrial activity in Ireland had not advanced to the second, science based phase, and also because of a lack of awareness by government.

Following the Meiji reforms, Japan turned to Europe for suitable models of educational institutions when making plans to catch up with western nations. A national college of technology was established in 1873 for the teaching of engineering, later absorbed into the Imperial University in Tokyo. During the 1880s when such hesitant steps were being taken in engineering education in Britain, every assistance was being given to expatriates who had been invited to work at the Imperial University of Tokyo. They were to establish a school of engineering, at the time the largest such institution anywhere. The reputation enjoyed by British engineering was no doubt the reason why so many of its staff were drawn initially from England, Scotland and Ireland. Among them was a graduate of TCD, C.D. West,[66] who remained in Japan until his death in 1908. Others included the Scotsman Thomas Alexander,[67] later to hold the chair of Civil Engineering at TCD, the graduate of the Queen's College Belfast, John Perry,[68] who was to teach at Finsbury College and another Scot, J.A. Ewing, later to hold the chair of mechanical engineering at Cambridge.

DECLINE OF OLDER INDUSTRIES

Irish manufacturing failed to develop any second-phase industries. One which seemed to have all the necessary prerequisites available to it was the manufacture of telescopes and optical equipment. Pioneering work was done in the technology of making giant metal mirrors by William Parsons, third earl of Rosse.[69] This was exploited by Thomas Grubb and his son, Howard. Prof. Humphrey Lloyd, who had played a decisive role in establishing the teaching of engineering at Trinity College had as a physicist, carried out research in optics, publishing a classical paper on conical refraction.

Technical Education and the Application of Technology

And yet the synergy which was to develop in German industries between manufacturer and academic, failed to emerge in Ireland. The firm, which later made periscopes for the Royal Navy, was moved to England during the first world war for security reasons. Eventually it was rescued by Charles Parsons, and as Grubb Parsons located at Newcastle upon Tyne, where it remained active in the manufacture of giant telescopes and infra red spectrometers until 1984. Clearly a number of factors besides technical competence are important for the success of such advanced industries.

The failure of industrialists in both islands to avail of men with the best technological education in their enterprises has played a significant part in the decline of indigenous industry. What might have been, is illustrated by the case of Charles Parsons.[70] The son of the third earl of Rosse and a graduate of Cambridge University, he established a successful business on Tyneside which exploited the latest science based technology. He pioneered one of the great inventions of the nineteenth century, the steam turbine and applied it commercially to the generation of electricity as well as revolutionising fast marine transport. As senior design staff he employed two Irish engineering graduates of Trinity College. In the 1890s his works management committee which met weekly, numbered two men who were to become members of the Royal Society. His firm prospered and survived well past the period with which we are concerned.

POLITICAL FORCES

Political independence in 1922 transferred power from a minority which shared many of the attitudes of the establishment in Britain, to the majority which had had a very different experience. An example of the consequences of this was the decision of the Irish Free State to organise a *national* electricity supply. Graduate engineers of all disciplines were given a key role and attention was focused on continental practice by the involvement of a German firm in the massive Shannon hydro-electric project. Among the population as a whole engineering in a broad sense still retained a positive image. But equally among the private individuals who owned industrial enterprises old perceptions were unchanged.

The attachment to education which the Catholic population demonstrated even under the Penal Laws, continued in the nineteenth century. But it was only when 'free secondary education' and adequate third level provision was made for technician training in the Republic, in the last twenty five years, that one could see the potential contribution to industrial development which their absence had precluded. The position was changed by a political decision of great consequences taken at the end of the fifties, just after the period being considered. It was decided to encourage multi-national firms to establish plants in Ireland with tax incentives. The result was a sharp influx of American, European and Japanese firms which had no inhibitions about employing graduates. The state too began to take a positive role encouraging the training of university graduates for employment in industry. The benefits which had been so clearly foreseen by such as Humphrey Lloyd in 1841, began to be realised.

Old influences still persist. Universities from their inception were established to meet the needs of society, educating future administrators, lawyers and doctors. To add another profession, namely engineering, would seem to be logical. However there has been a price to pay. The universities preserve the value systems and attitudes of society. Many among the newly empowered majority population exhibited the conservatism of a rural population. Farmers after all must run businesses which, if mismanaged will bring ruin. The universities have also inevitably absorbed some of the attitudes towards industry which characterised the wealthy and influential members of society in Britain. Such attitudes have become more resistant to change because they have helped to determine the behaviour of some future civil servants and even of engineering graduates long after power has passed from the old establishment which engendered them. Present day efforts to develop indigenous industry have to contend with this relic.

Conclusion

One and a half centuries is a period long enough to witness major changes. Developments in technology have transformed many manufacturing processes from a craft base to a science base. During this period literacy has increased enormously, and the teaching of science and technology has been introduced at all levels into the educational system. Political action in the form of the Act of Union reinforced the influence of England on developments. Nevertheless during the whole period there was an awareness of how education was evolving in other European countries. One channel for such communication was Dublin University which enjoyed something of a golden age academically in the last century. Because in Ireland the state had greater ability to implement its wishes, and because there were citizens with clear ideas for the future who could influence decision making, developments in technical education in Ireland differed significantly from the experience in England. The political transition in 1922 which marked the end of the Union, has opened the way for developments allowing full advantage to be taken of such differences, although the full effects have only become clear since 1950.

Notes

1. R. Delaney, *Ireland's Inland Waterways* (Belfast, 1992).
2. R. Preece, 'The Durham Students of 1838', *Transactions of the Architectural and Archaeological Society of Durham and Northumberland*, 6 (1982): pp. 71–74.
3. L.M. Cullen, 'Population Growth and Diet', in M. Goldstrom and L. Clarkson, eds., *Irish Population, Economy and Society* (Oxford, 1981), pp. 81–92, see p. 111.
4. J. Herivel, 'Carnot and the French Scientific Milieu around 1824' in Centre National de la Recherche Scientifique, *Sadi Carnot et l'essor de la thermodynamique* (Paris, 1976), pp. 81–92.

5. R.R. Locke, *The End of the Practical Man: Entrepreneurship and Higher Education in Germany, France and Great Britain, 1880–1940* (London, 1984), p. 31.
6. M. Crosland, *Science under Control, The French Academy of Sciences 1795–1914* (Cambridge, 1992), Chapter 1.
7. *Ibid.*, p. 43.
8. Santiago Riera i Tuèbols, 'Industrialization and Technical Education in Spain 1850–1914', in R. Fox and A. Guagnini, eds., *Education, Technology and Industrial Performance in Europe 1850–1939* (Cambridge, 1993), pp. 141–170.
9. J.C. Baudet, 'The training of engineers in Belgium, 1830–1940' in Fox and Guagnini, eds., *Education, Technology and Industrial Performance*, pp. 93–114.
10. A. Guagnini, 'Academic Qualifications and Professional Functions in the Development of the Italian Engineering Schools, 1859–1914' in Fox and Guagnini, eds., *Education, Technology and Industrial Performance*, pp. 171–195.
11. Locke, *The End of the Practical Man*, p. 58.
12. Herivel, 'Carnot and the French Scientific Milieu around 1824', p. 88.
13. A. Guagnini, 'Worlds apart: Academic Instruction and Professional Qualifications in the Training of Mechanical Engineers in England, 1850–1914', in Fox and Guagnini, eds., *Education, Technology and Industrial Performance*, pp. 16–41.
14. Report from the Parliamentary Select Committee of Inquiry into Instruction in Theoretical and Applied Science to the Industrial Classes (London, 1867–68).
15. Locke, *The End of the Practical Man*, p. 49.
16. Quoted in W.H.G. Armytage, *A Social History of Engineering* (London, 1971), p. 122.
17. Guagnini, 'Worlds apart', p. 24.
18. P.J. Dowling, *A History of Irish Education* (Cork, 1971), pp. 106–114.
19. K. Byrne, 'Technical Education', in A. Hyland and K. Milne, eds., *Irish Educational Documents* (Dublin, 1987), pp. 239–301.
20. F. Brennan, ed., *Kevin Street College 1887–1987* (Dublin, 1987), p. 7.
21. Dowling, *A History of Irish Education*, pp. 139–158.
22. T. Corcoran, ed., *The National University Handbook* (Dublin, 1932); see 'Statistics of University Entrance', pp. 271–272.
23. L.M. Cullen, *An Economic History of Ireland* (Oxford, 1981), pp. 77–99.
24. Byrne, 'Technical Education'.
25. T.W. Moody and J.C. Beckett, *Queen's Belfast 1845–1949: The History of a University* (London, 1959), pp. xliv–liii.
26. *Ibid.*, p. xlviii.
27. M.T. Casey 'Nicholas Callan: Physicist' in C. Mollan, W. Davis and B. Finucane, eds., *Some People and Places in Irish Science and Technology* (Dublin, 1985), pp. 30–31.
28. R.B. McDowell and D.A. Webb, *Trinity College Dublin, 1592–1952* (Cambridge, 1982), p. 159.
29. R. Kane, *The Industrial Resources of Ireland* (Dublin, 1845).
30. Report from the Parliamentary Select Committee of Inquiry into Theoretical and Applied Science to the Industrial Classes, para 2841.
31. See the exhibition catalogue by R.C. Cox, *Engineering Ireland 1778–1878* (Dublin, 1978).
32. Moody and Beckett, *Queen's Belfast*, pp. liii–lxvii.
33. McDowell and Webb, *Trinity College*, p. 159.
34. Report from the Parliamentary Select Committee of Inquiry into Instruction in Theoretical and Applied Science to the Industrial Classes, para 2841.
35. H. Lloyd to E. Sabine, 1841; Sabine papers, BJ/3 10–11, Public Records Office, London.
36. H. Lloyd, *Praelection on the Studies connected with The School of Engineering* (Dublin, 1841), p. 22.

37. *Ibid.*, p. 26.
38. *Ibid.*, p. 29.
39. J.G. Byrne, R.C. Cox and W.G. Scaife, '150 Years of Engineering at TCD', *Engineer's Journal*, 45 (1992): pp. 36–39 and 52–59.
40. R.V. Dixon, *Scientific Training for Practical Pursuits, a Farewell Address to the Students of the School of Engineering, Trinity College Dublin* (Dublin, 1854).
41. Byrne et al., '150 Years of Engineering at TCD'.
42. Moody and Beckett, *Queen's Belfast*, p. 69; also P. Leahy, 'History of Engineering Training in Ireland', *Engineer's Journal*, 15 (1962): pp. 147–51 and 208–211.
43. Moody and Beckett, *Queen's Belfast*, p. 286, and Leahy, 'History of Engineering Training in Ireland'.
44. Leahy, 'History of Engineering Training in Ireland'.
45. See A.R. Hall, *Science for Industry – A Short History of Imperial College* (London, 1982), p. 24, and Report of Select Committee, p. 25.
46. K. Gispen, *New Profession, Old Order: Engineers and German Society, 1815–1914* (Cambridge, 1989), p. 73.
47. Locke, *The End of the Practical Man*, pp. 32–35.
48. Byrne et al., '150 Years of Engineering at TCD'.
49. Evidence given to the Royal Commission appointed to 'Inquire into the State . . . of the University of Dublin', House of Commons Parliamentary Papers, 1852–53, vol. XLV, p. 345.
50. R.C. Cox, *Engineering at Trinity* (Dublin, 1993), p. 39 and p 49.
51. Byrne et al, '150 Years of Engineering at TCD' and Leahy 'History of Engineering Training in Ireland'.
52. W.E. Coe, *The Engineering Industry of Northern Ireland* (Newton Abbot, 1969), p. 173.
53. See J. Bardon, *Belfast: An Illustrated History* (Belfast, 1982) and M. Moss and J.R. Hume, *Shipbuilders to the World: 125 Years of Harland and Wolff* (Belfast, 1986).
54. Locke, *The End of the Practical Man*, p. 58.
55. *Ibid.*, especially chapter 2, 'The Graduate Engineer and Industrial Performance'.
56. R.E. Jacob, 'Quakers in Industry and Engineering in Ireland in the Nineteenth Century', *Transactions of the Institution of Engineers of Ireland*, 112 (1988): p. 461.
57. Guagnini, 'Worlds apart', p. 22.
58. Hall, *Science for Industry*, p. 18.
59. J.F. Gillies and O.M. White, *Technical Education in Northern Ireland* (unpublished memoir, University of Ulster).
60. Bardon, *Belfast: An Illustrated History*, p. 209.
61. W.C. Borrowman, 'Some Considerations Concerning the Training of Young Engineers', *Transactions of the N.E. Coast Institute of Engineers and Shipbuilders*, 16 (1900): pp. 113–143.
62. Moody and Beckett, *Queen's Belfast*, and Leahy, 'History of Engineering Training in Ireland'.
63. McDowell and Webb, *Trinity College*, p. 406.
64. Hall, *Science for Industry*.
65. Locke, *The End of the Practical Man*, p. 34.
66. Kaoru Hongo, 'The Father of Japan's Mechanical Engineering', *Look Japan*, 31 No. 349 (1985): p. 22.
67. Cox, *Engineering at Trinity*, p. 59.
68. W.H. Brock, 'Reforming Mathematics Education: an Anglo-Irish Connection', in J.R. Nudds et al., eds., *Science in Ireland, 1800–1930* (Dublin, 1988), pp. 31–49, see p. 36.
69. W.G. Scaife, 'From Galaxies to Turbines', *Physics Bulletin*, 39 (1988): pp. 103–105.
70. W.G. Scaife, 'Charles Parsons – Manufacturer', *J. Mat. Processing*, 33 (1992): pp. 323–330.

Chapter 7

Some Aspects of the Evolution of Agricultural and Technical Education in Nineteenth-Century Ireland

Richard A. Jarrell

INTRODUCTION

The nineteenth century was a period obsessed with technical education. The triumphs of the physical sciences during the seventeenth and eighteenth centuries encouraged the belief, held by philosopher and industrialist, as well as agriculturist and politician, that the application of science would create great wealth. For them, education was the essential conduit between scientific theory and practice. If only scientifically untutored people could understand the working of the world, economic improvement would follow. I take 'technical education' in its broadest sense: education aimed at improving the practice of industry, whether artisanal, craft or factory, and of agriculture. In the British Isles, during the late eighteenth century, the vision of technical education played itself out not in formal educational structures but in a variety of indirect ways, primarily through societies. The organisers were drawn from the upper and middle classes: clergymen, academic scientists, politicians, landowners, merchants, industrialists, professional men and farmers of means.

When the nineteenth century dawned, the paths chosen by Great Britain and by Ireland to prosecute the goals of technical education, despite the Union forged in 1801, diverged sharply. In Britain, the *laissez-faire* attitude towards industry and education was only reluctantly, and not fully, abandoned after three-quarters of a century of struggle and experience. The Irish, on the other hand, embraced state intervention from the first. Even when challenged and defeated by British interests later in the century, they persevered and returned to their preferred solutions by century's end.

This study calls attention to the parallels between Irish agricultural and technical education and argues how the two, though having different aims, shared essentially the same characteristics. Because of sizeable state support for these ventures, Treasury and parliamentary select committees and Royal Commissions scrutinised every step in microscopic detail. Thus, we have a wealth of information about the workings of Irish technical education. Particularly useful is the Irish evidence taken in 1883 by the Royal Commission on Technical Instruction, chaired by Bernhard Samuelson. It provides us with a 'snapshot' of the Irish situation and attitudes before the separate Irish Department of Agriculture and Technical Instruction appeared in 1900.

However, we cannot take the history of technical instruction out of context. The stages I will suggest, and their characteristics, must be seen against the backdrop of economic, social and intellectual history. Thus, the kinds of activities we meet during each phase reflect the relative state of development of Irish industry and agriculture, along with the social response to it. The intellectual response at each stage – and I will argue that much of it was borrowed from elsewhere – likewise reflected the reaction to the economic and social conditions as the elite perceived them at the time.

By way of providing economic and social signposts, let me first sketch a vastly oversimplified summary of the Irish industrial and agricultural economy of the nineteenth century.[1] Beginning before the Union in 1801, Irish agriculture and industry had prospered, thanks to Britain's wars. Exports thrived and farmers did well, though not through greater production but rather from inflationary prices. This prosperity collapsed in 1815 and was soon replaced by rapidly falling prices for agricultural produce, a banking crisis and several years of poor harvests and famines. During the 1820s and 1830s, agricultural prospects improved, although agricultural labourers continued to fare badly. Small farmers, who could supplement their incomes through textile production, could be comfortable. Before the Famine, the rural population was not homogeneous: independent farmers and tenants with larger holdings, having some capital, marrying later, subdividing their land less and eating a mixed diet, were concentrated in eastern Leinster and eastern Ulster. The less well-off were found in parts of central and western Ireland, with the poorest in the far west. The agricultural labourers, with their excessive land subdivision, potato-dominated diet, early marriage and high birth rate, were by far the most disadvantaged and they bore the brunt of the agricultural disaster after 1845, although famine and want were their constant companions well before then.

The woollen and cotton industries, already in decline due to competition from Yorkshire and Lancashire, faded after the depression of 1825–26. However, the linen industry continued to grow and centralise in the Belfast area. Malting and brewing, moving from an artisanal to factory basis, was on the rise by the 1830s. On the eve of the Famine, despite the successes of the linen industry, Irish industrialisation was weakening and this at a time when the railway boom was underway. The Famine seems to have little impact upon industry, which recovered and grew until the end of the 1870s. Nor did the Famine uniformly impinge upon the rural population. The better-off farmers escaped much of the disaster. Those who did not, emigrated, allowing those who remained to increase their holdings. Ironically, emigration was greater in eastern Ireland than the poor west, where earlier social and economic practices survived. The net result for much of the country was the near disappearance of labourers and cottiers. Before 1880, with good harvests, rising prices, moderate rent increases, rising wages due to a labour shortage and a move to cattle raising and dairying, prosperity returned to the countryside.

The period from 1879 to about 1890 saw a serious deterioration in both industry and agriculture. The industrial crisis in England led to the flooding of the Irish market with cheap goods, poor harvests led to a serious dislocation in farming and heightened political friction. By the last decade, however, with land reform, rising prices, continued emigration and social change in the west, agriculture rebounded. Alongside it, industry expanded. By the turn of the century, about one-third of Irish exports were industrial products, concentrated upon linen, beverages and shipbuilding and engineering.

AGRICULTURAL EDUCATION IN IRELAND

Amongst the set of paradoxes that was nineteenth-century Ireland, perhaps none is so rich for the historian as agriculture. Ireland in 1800 was a small island with a rapidly-growing population and a social system that encompassed magnates with vast estates, small independent farmers and a peasantry perhaps no better off than Russian serfs.[2] Despite its wide variety of landscapes and soils and a moderate climate, only three principal crops – wheat, flax and potatoes – dominated its agriculture. The failure of the last led to a social and economic dislocation almost unparalleled in modern European history, the effects of which are still evident a century and a half later. Ireland was blessed and cursed by agricultural riches – potentially blessed and actually cursed. I wish to focus upon only one small part of that story, the drive to secure for the wider rural population expertise in the best methods of farming. This story, too, is paradoxical, in that Ireland was perhaps the world leader in agricultural education during the last century, but in large part was unable to capitalise on its innovations.

Efforts to improve farming were scarcely novel by 1800. Members of the Dublin Philosophical Society at the turn of the eighteenth century discoursed upon the subject. The Dublin Society, later the Royal Dublin

Society (RDS), founded in 1731, made some desultory efforts to improve farming.[3] For example, an essay written for a competition in 1792 extolled the virtues of creating a model farm within a wider scheme of national education. Local agricultural societies also date from this period, leading to the formation of the Farming Society of Ireland in 1800. The Cork Institution (1802) and the Agricultural Improvement Society (1833), along with the RDS, provided national and regional forums. This movement was not unique to Ireland, of course. The English Agricultural Society and the Scottish and Highland Society were active contemporaries. Such societies catered to the upper class and prosperous farmers and focussed upon livestock breeding and agricultural shows.

Modern Irish agriculture is efficient and progressive. Science was a key factor in its maturation, but that is a twentieth-century story. I would suggest that efforts to transform Irish agriculture with scientific knowledge in the nineteenth century were largely failures. One can identify several distinct phases in the evolution of Irish agricultural education in the last century, each succeeding phase more greatly emphasising the role of science. I will sketch these to provide a framework, but will concentrate upon more general questions regarding the economic role of agricultural education.[4]

1. *The first phase*, lasting into the early 1830s, is a collage of small efforts and much talk. Local societies along with forward-looking landowners and their managers, introduced improved methods from Scotland and England, mainly into Ulster. A quasi-apprentice system sprang up, with local boys working on farms under the tutelage of managers such as William Blacker, estate agent for the Earl of Gosford in Co. Armagh. Two notable schools, 'Martin Doyle's' in Bannow, Co. Wexford, and the Templemoyle School near Derry, organised in the 1820s and 1830s, attempted similar instruction to small classes of local boys. Their curricula combined theoretical (and probably useless) science with practical techniques. Local agricultural societies sprouted up just after the turn of the century, only to go into decline by the 1830s.

The leading agricultural writers of the time, including Doyle, Edmund Murphy and John Kelly, dismissed the waste of large government grants to societies and pressed for local schools to train sons of farmers in the latest techniques. These techniques were not so much scientific as technological: the use of new tools, the introduction of a wider variety of crops, rotation, drainage and better animal husbandry methods.[5] The obstacles to their adoption were formidable. There was little incentive for the landowner, if his rental income was sufficient, and even less for the poorest tenant, who might believe that any improved productivity on his part would lead to higher rents. These obstacles never entirely disappeared. By 1870, when the economic state of farming was reasonably good thanks to good harvests and higher prices, J.N. Murphy noted the practice of landlords as a rule, on the determination of a tenancy, demanding the full improved letting value of the land. This practice operates as an effectual bar on improvements, as tenants will not freely invest their labour and capital at the risk of the proceeds going into the landlord's pockets instead of their own.[6]

The landlords, themselves, might have been the vectors for improved methods – as many were in England – by introducing new techniques in tandem with capital improvements. However, as Vaughan shows,[7] Irish landlords invested little in improvements during the century and the trained agriculturist on the landlord's payroll remained a rare occurrence.

From an economic perspective, this period was generally one of relative agricultural prosperity, except for the years immediately following the Napoleonic Wars. The emphasis of educational proponents was on individual initiative. Efforts were small in scale and imitative in nature, but if a marginal farmer paid attention to the lessons of more prosperous neighbours, he might improve his lot.

2. *The second phase* began with institutionalised training when the Board of National Education introduced agricultural instruction. This is a well-known story and few details are necessary here.[8] The jewel in the crown of the system was the Glasnevin Model Farm, established in 1838, to provide agricultural training for student teachers at the Marlborough Street institution. In time, many model agricultural schools appeared throughout the country, along with national and private schools with small farms attached to them. In the end, free-trade politics brought them down and by the mid-1870s, all had disappeared but the Glasnevin and Cork schools. No full study of these schools exists; it is questionable whether they had much real influence in Ireland. Nonetheless, they do represent perhaps the most comprehensive state-supported system of agricultural instruction the world had seen to that time. The system came into being at a time of relative agricultural prosperity but, probably more significantly, it was a bureaucratic invention, an extension of a state-controlled educational programme.[9]

3. *The third phase*, the introduction of agricultural education into the universities, was an attempt to emulate, at a higher level, the efforts of the National Schools. The introduction of the Queen's colleges in the 1840s aimed to provide non-sectarian higher education to a broader populace. Agricultural instruction in the colleges led as chequered a life as the colleges themselves. Dr John Hodges, a Belfast physician and improver, introduced a full programme at Queen's, Belfast, in 1851, while Edmund Murphy held the chair at Cork and Thomas Skilling at Galway. Between 1851 and 1862, the three colleges had awarded certificates to a total of fourteen students. By the mid-1860s, such instruction had virtually disappeared. The Royal College of Science for Ireland, established in 1867, taught agricultural chemistry until it abolished the agricultural chair in 1878.

How did university-based instruction differ from that of lower schools? In the first place, agriculture was supplementary to a normal collegiate course of studies. Its content was theoretical and textbook-based, emphasising agricultural chemistry, mensuration, farm management and economics. No college had a suitable model farm. Testimony of professors before commissions of enquiry lamented these facts, along with the problem of rural boys lacking a proper education to take full advantage of the academic instruction. Detractors of university instruction argued that agriculture was best taught locally, through the National Schools, centring upon practical training on farms attached to the schools. The timing of this phase is important, as it was an early response to the Famine. This response implied that the training of an agriculture elite to become agents of change would have help to forestall future disasters.

4. *The fourth phase*, covering the last two decades of the century, is a mixture of parallel efforts that came closer to success and laid the foundation for twentieth-century improvement. The result was the confluence of several streams: the cooperative movement of Horace Plunkett and the Irish Agricultural Organisation Society (IAOS),[10] the crescendo of Home Rule agitation, efforts of the Congested Districts Board, greater interest in technical and secondary instruction generally and the realisation that agricultural progress was a form of industrial progress. By this time, formal agricultural education was only a shadow of its former self. Only two model schools survived after 1873, those in Glasnevin and Cork. The latter became the Munster Institute in 1880, capitalising on the south-west's long tradition of dairying and butter-making, becoming the centre for modern dairy training. It expanded into poultry and other areas of animal husbandry in the 1890s.[11]

The cooperative movement and the development of creameries underscored the fact that agriculture had become a food industry. Writers of the 1880s and 1890s who pressed for agricultural instruction argued that Ireland needed to harness science and technology to forge ahead economically. Key Home Rulers in Westminster believed salvation lay in that direction, and that only a Dublin-based bureaucracy could foster progress. Through the untiring efforts of Plunkett, by then a Unionist MP, and the Recess Committee, this vision came to fruition with the passage of the Agriculture and Technical Instruction (Ireland) Act in 1899. The critical state of Irish agriculture in the late 1870s and 1880s was a stimulus to change. Interestingly, despite serious social and political unrest, important efforts for education were made by associations, like the IAOS and the RDS, by energetic individuals and institutions. Government, on the other hand, almost completely retreated from the field. The return of the state, the preferred solution of many Irish commentators, came only when an Irish department emerged.

Technical Education in Ireland

If the history of Irish agricultural education is paradoxical, the history of technical education is incongruous. In Britain, technical education developed in response to industrialisation, partly to maintain a competent workforce, and later, to attempt to catch up with rivals such as Germany, France and the United States. Ireland was seriously under-industrialised in the nineteenth century, a victim of British actions and its own inability to generate industrial capital. As in agricultural education, the Irish took their initial cues from Britain.

Again, one can conveniently divide the history of technical education into four phases, which in temporal terms, roughly parallel those of agriculture.

1. *The first phase* lasted until the mid-1840s and was characterised by the activities of societies and proprietary institutions. The Royal Dublin Society was the central organisation. Relieved of agricultural matters by the Farming Society in 1800 and assured of continuing substantial parliamentary support, the RDS turned its efforts towards manufactures and natural resources. Its formal educational programme was limited to its Dublin drawing school, but its lectures offered the public information on science and industry. This was equally true of provincial institutions in Cork and Belfast and especially of the mechanics' institutes that appeared on the scene in the 1830s. It is likely that such institutions influenced very few 'mechanics and artisans'. Compared with Britain, where mechanics' institutes numbered in the hundreds and could in theory reach tens of thousands of industrial workers, only a few institutes operated in Ireland; outside the three cities, institutes existed only in Birr, Dundalk, Galway, Limerick, Waterford and Downpatrick. Industrial workers, concentrated in Dublin, Belfast and Cork, represented a small portion of the nation's working population. The Dublin Mechanics' Institute enrolled nearly 3500 members by 1852, but its classroom programme reached only 176 students, its lone technical course (drawing) eclipsed by more popular offerings in English, French, Italian, dancing, vocal and instrumental music. Given its activities, one would be dubious of its supposed working-class orientation.

2. *The second phase* began just before the Famine and marks the debut of state-funded institutions and programmes. In 1845, Robert (later Sir Robert) Kane founded the Museum of Irish Industry in St Stephen's Green, modelled on the Museum of Economic Geology in London and using the collection of the Irish Geological Survey as its core.[12] In his inaugural lectures, published as *The Industrial Resources of Ireland*, he attempted to stimulate interest in Irish manufactures. Ireland's problem, as he saw it, was not the lack of natural resources, labour or innate genius but the lack of education. Thus, one thrust of the new Museum's activities was the provision of courses of lectures on scientific and technical topics. The success

of the Crystal Palace exhibition of 1851 led to a similar, though smaller, exhibition in Dublin in 1853.

This period also saw a transition to almost complete London-controlled science and art training. The government created the Department of Science and Art in 1853 as one response to the educational challenge suggested by the Great Exhibition. As Britain lacked a true education department – and England a state educational *system* – the Science and Art Department devised and directed technical instruction, mostly in evening schools, for nearly a half-century. Henry Cole and Lyon Playfair, the founding secretaries, had a simple vision: teach working-class youth basic science, mathematics, drawing and art to enhance their abilities to participate in industrial innovation and design. By 1859, the department prepared an examination scheme, implementing the 'payment-by-results' system of funding teachers.[13] The system was slow to develop in Britain, and even slower in Ireland, as both lacked trained teachers and widespread science-teaching facilities. During the early 1860s, the only means to bring teaching to Irish students was through provincial courses of lectures, given by government-supported itinerant lecturers. Irish critics argued that the department's scheme suited the English working class, a class almost absent in Ireland.[14]

3. *The third phase* saw the creation of new tertiary-level scientific and engineering education.[15] In the Queen's colleges, science and engineering fared better than agricultural instruction, but due to the opposition of the Roman Catholic hierarchy to nondenominational higher education, few Catholics attended. The only alternative, Trinity College, Dublin, which had a long tradition of excellence in both areas, attracted few Catholics or non-Anglican Protestants. Newman's Catholic University of Ireland, founded in 1852, included a Faculty of Science – W.K. Sullivan was its Dean – but it attracted few students. Given the mixed success of the Queen's colleges, the state introduced another institution in 1867, converting the Museum of Irish Industry into the Royal College of Science for Ireland.[16] The college had a first-rate staff in pure and applied science and ample scholarships. It had its modest successes but attracted about half its student body from Britain, where similar facilities were lacking. More discouraging was the fact that many well-qualified graduates of Irish third-level institutions emigrated for lack of opportunity at home.[17]

4. *The fourth phase* covers the last two decades of the century. If we can characterise the second phase as a 'solution in search of a problem', the fourth is the reverse. In Britain, by the early 1880s, widespread concern about technical education and the weaknesses of the Department of Science and Art system led to the appointment of the Royal Commission on Technical Instruction. It recommended the establishment of local educational authorities to provide solutions. In Britain, the Local Government Act (1888), the Technical Instruction Act (1889) and the Local Taxation Act (1890) allowed for, respectively, the creation of county councils,

giving them the power to build, direct and finance technical schools out of the rates, and providing them with revenue. After that, the Science and Art examination system blossomed, along with a host of local technical schools. In Ireland, progress was much slower, with no local government act until 1898. By 1895, while English science schools enrolled 145,000 students, Irish schools taught only 6500 (fewer than Wales).[18] Still, energetic individual efforts resulted in the appearance of technical schools in Belfast, Dublin and other cities.[19] An important feature of the educational thrust of the new technical schools was a move away from the tradition of pure science teaching to more practical instruction.

Although technical and agricultural education followed parallel but separate paths, they finally converged, at least administratively, in 1900 with the formation of the Irish Department of Agriculture and Technical Instruction under Plunkett and T.P. Gill. All Irish operations of the Science and Art department transferred to Dublin; these included the Royal College of Science, the Irish Geological Survey, the National Museum, the National Gallery, grants to institutions and the Science and Art examination and grants system.[20]

Lacking details of technical instruction for this period, we can only speculate on the motivations of those who participated. The examination scheme of the Department of Science and Art targeted the 'working or industrial classes', though that meant in practice its upper echelons, sons of skilled tradesmen and semiskilled operatives, not the working poor. This may also have prevailed in Ireland. The syllabuses in authorised science classes and schools and the examinations concentrated on pure, didactic science, not on practical information or manual skills. In England, where secondary education was virtually nonexistent for the mass of working people, the Science and Art examinations were the only means for post-elementary education. In time, lower middle class youth availed themselves of Science and Art instruction to 'get on'. As Ireland had no system of mass secondary education,[21] its youth, too, may have seen in technical instruction a way to pursue further education.

IRISH TECHNICAL INSTRUCTION: THE SITUATION IN 1883 AND AFTER

By the early 1880s, British concerns about the efficacy of existing technical instruction called into being another Royal Commission. Ireland was not an import focus of this enquiry, and the commission's report led to more immediate results in Great Britain. However, the commission's work does seem to have been a spur to local Irish activity, which did lead to independent action.

The evening science classes to prepare students for Science and Art Department examinations had had little success in Ireland. Table 1 shows the number of science and art students in the United Kingdom in 1883–84.

The popularity of these programmes was considerably greater in Scotland, despite its smaller population, than in Ireland. A closer look at the location of science 'schools' – they might only have been single classes – reveals that most students were drawn from the three cities, Dublin (420), Belfast (597) and Cork (188). Ulster also had the most schools (98), followed by Munster (29), Connacht (15) and Leinster (10). Nearly 3000 students (63%) were located in Ulster. While it would be tempting to explain this disparity by appealing to cultural differences, it is more likely that the key factor was the level of industrialisation. By 1894–95, Dublin enrolled nearly twice as many students as Belfast, with the greater Dublin region and Ulster each accounting for about a quarter of the Irish total.[22]

Table 1. Science and art enrolments 1883–84

Country	In Science Schools	In Art Schools
England	56,777	47,964
Wales	1,867	870
Scotland	11,630	7,221
Ireland	4,619	1,816
Total	**74,893**	**57,871**

(Department of Science and Art, 36th Annual Report, 1885)

As background, we should note that Irish under-industrialisation had two particular structural problems. First, the proportion of the population directly involved in industrial pursuits was smaller than in Great Britain. Under the 'industrial' head of census-defined occupations in 1871, Ireland counted 10% of its productive population, compared with 23% each in England/Wales and Scotland; this rose to 13% by 1881. Agriculture, on the other hand, accounted for 20% of the Irish productive population in 1871, compared with 7% in England/Wales and 8% in Scotland. The agricultural proportion in Ireland dropped to 19% in the ensuing decade.[23] But general statistics mask the second structural defect, the nature of industrial enterprises. Irish industry was, for the most part, smaller in scale. It did not encompass iron and steel production, textiles (save for the failing linen industry), coal mining, machine tools, large-scale railway equipment or agricultural implement manufacturing. Thus, a juxtaposition of the content of Irish technical instruction with the existing industrial base, then it would appear deficient in the eyes of outside observers.

Evidence taken by the Samuelson Commission in Ireland during 1883[24] shows us not only the state of instruction, but also the differing ideological stances of those interested in pressing for such education.

The commissioners – none of them Irish – took evidence only in Belfast, Dublin and Cork. Differences of opinion between the academics and manufacturers were striking: almost to the man (except Sir Patrick Keenan, Resident Commissioner of National Education), the academics favoured the continuing teaching of scientific principles. The manufacturers supported a variety of more practical training, though there was no unanimity amongst them. Sir John Preston offered the extreme position: no education would help the linen industry's competitiveness, only the introduction of a sixty-hour week. Most witnesses, representing small-scale industries and trades, offered a litany of Irish failures to keep abreast of new techniques or to learn useful information relating to their trades.

The picture painted by the witnesses, and reported by the commissioners, was that Ireland possessed sufficient natural resources (as Kane had argued nearly forty years earlier), but that its agriculture was backward, its traditional industries such as furniture-making were in decline and its linen trade no longer competitive. Irish manufactures exhibited a poverty of design. With such a set of problems, it is no wonder that those appearing before the commission offered such disparate solutions. William K. Sullivan, by then president of Queen's College, Cork, outlined the most comprehensive scheme to the commissioners. Sullivan, an admirer of Prussian technical education (he held a German Ph.D. in chemistry), argued for a hierarchical system of local, provincial and national art schools, along with improving local science schools, expanding the Royal College of Science and converting the Queen's Colleges into polytechnics.[25] The witnesses were in accord that Irish manufacturers were far behind Britain, and they clearly thought in terms of building a similar industrial base in Ireland.

In its report, the commission separated its recommendations for Great Britain from those of Ireland (set aside as 'special recommendations').[26] Rather than proposing anything radical, the Commissioners simply selected some non-controversial suggestions made to them during their hearings. These included compulsory, primary education, systematic manual training and a payment-by-results scheme for local industrial teaching and Treasury grants to local agricultural and industrial schools. The commissioners assumed that local initiative was appropriate for Britain, but that they could not ignore the long-entrenched Irish view of state assistance: 'There is a general disposition to look to the Government for the initiative, and a conviction that local efforts must be sustained by State subsidies.'[27]

In keeping with the ideal of science teaching as the basis for technical instruction, they followed the academics' advice concerning evening science tuition by the Royal College of Science and for improved secondary-level science education. It was not without condescension that the commissioners recommended that the Irish concentrate upon local manufactures such as lace-making, basket-weaving, woodcarving and the production of small items like gloves and straw envelopes for bottles.

Despite the commissioners' special recommendations, the government had no intention of including Ireland in a general scheme for technical instruction. Lord Monteagle, rising in the Lords in July 1887, asked whether the proposed technical education bill would include Ireland and whether it would address agricultural instruction. Viscount Cranbrook replied that

> ... the Bill which had been proposed was intended only for England but the principle extended to Ireland, though the form of the measure might be unsuitable for that country, nor did he think that Ireland could be satisfactorily included in the Bill.[28]

Nor did the government intend to include agriculture in its bill. The Irish problem was the lack of local authorities and debate in the Commons over the next months focussed upon the need for local support, not state funding. So local initiative had to suffice. In 1887 a private group, armed with a city corporation grant of £500, opened the City of Dublin Technical School in Kevin Street.[29] Progress remained slow for the next decade. Only eleven municipalities made use of the 1889 act's provisions.[30] The creation of local councils for Ireland had no immediate effect. T. Preston, a Science and Art Department inspector complained in 1899 that grants had fallen and schools had declined and that the decline was '... likely to continue under existing conditions until science teaching becomes practically extinct in Irish schools – a point, indeed, which is now very closely approached'.[31] The problem, he believed, was the structure of Irish education and that

> The Local Authorities in Ireland have no experience (speaking generally) in educational matters, and it is not likely that they will give aid towards, or take any interest in, any educational system. It is for this reason that the Technical Instruction Act remains almost entirely inoperative throughout the country.[32]

In the end, only a form of 'home rule' for Irish technical education and agricultural instruction found favour and even when a new system emerged, the battle for technical instruction would continue for another thirty years.

SOME GENERAL REMARKS

From even a casual reading of the history of Irish agricultural and technical education, one is struck by the sheer amount of activity. In agriculture, the widest-ranging efforts were those of the Board of National Education, but politicians who doubted their efficiency curtailed them. Irish agriculture did become more efficient and productive, but the extent to which agricultural instruction played an important role is debatable. Indeed, at the time the most National School agricultural graduates were active and might have made a difference, worldwide depression and changes in the British market reduced the value of Irish farm production drastically. Rural violence and the continuing loss of population exacerbated the problem. Agricultural instruction may have made a difference – we do not know at this stage of scholarship – but larger events masked such gains. Much of the emigration during the second half of the century seems to have better-off and better-educated farmers. How many took their National School skills elsewhere?

The state of agriculture across Ireland in 1900 is striking in its diversity. The farmers of Wexford, for example, were quite progressive, and this is reflected by the buoyant agricultural implement industry in Wexford city. This was also true of certain parts of Leinster and in eastern Ulster and, of course, of the dairy industry in Munster. In the west, however, the area targeted by the Congested Districts Board, a combination of poor land and social impediments conspired to suppress any real progress. This diversity needs explanation. One suspects that social factors were perhaps more important than economic or geographic factors in favouring certain regions.

Ireland's industrial underdevelopment explains why technical education was immature compared with Britain and the Continent. Yet, throughout the century individual writers, societies and governments touted technical instruction as the panacea for Ireland's economic and social woes. Again, this was not unique to Ireland, but a worldwide phenomenon. Ireland generally took its models from across the Irish Sea. This movement had spawned mechanics' institutes, state-supported lecture series, local and national societies and institutions such as the Cork Institution, the Museum of Irish Industry, the Royal College of Science for Ireland and, eventually, local science and art schools. Given industrial underdevelopment, why so much activity? Partly, I believe it arose from a naive belief in the power of science to transform production and labour, yet the efficacy of any of these forms was questionable in the Irish context and had loud detractors as well as ardent promoters.

An examination of activities in both agriculture and technical instruction, however, shows that the interest level did not remain constant. It was strong at the beginning of the century, when national institutions (such as the RDS), local institutions and agricultural societies flourished. One suspects that this reflected the late eighteenth-century faith in science. Activity increases again in the late 1840s and early 1850s, partly due to the enthusiasm sparked by the international exhibition movement and partly as a search for scientific solutions to Famine-induced social dislocation. By the 1870s, activity wound down as political and economic turmoil displaced science and technology. The last two decades see a subtle shift to a different vision of the delivery and content of agricultural and technical education. Old solutions had failed and new ones had yet to emerge. I would suggest that, in part, this ebb-and-flow, or wavelike phenomenon, was generational. The leaders first attracted to science in the 1780s and 1790s dominated the first decade of the century. A different generation – the Kanes, Sullivans and Hodges – was the mature influence in the middle of the century, but by the 1890s, they were all retired or dead. Those who led the way at century's end, the Plunketts and Gills, grew up in a different political, social and economic environment and in a time when the sophistication of science distanced it from ordinary education.

I would also suggest that two fundamental flaws in the technical education movement – and more specifically, the agricultural education movement – hampered any efforts to improve industrial or agricultural productivity in the nineteenth century. One was the political nature of these movements: they were typically 'top-down', i.e., organised by and ultimately for those who controlled industry and agriculture. People at the grass roots almost never organised such education. This was true even of the cooperative movement of the 1890s.

The other flaw was the definition of 'proper' technical instruction. A cursory survey of the history of technical instruction easily reveals the narrowness of the curriculum and methods employed once such education became scientific. Instruction for industrial workers concentrated upon theoretical scientific knowledge or technical skills such as drawing; agricultural instruction focussed upon topics such as soils and organic chemistry. Many writers insisted on the uselessness of teaching agriculture (or any trade) from textbooks.

Technical education is also problematical in terms of its appropriateness. One can show in many cases during the last century that the methods taught were either obsolete or inappropriate, or both (one might argue this is still true). Combined with the inertia typical of educational systems, technical instruction did not immediately reflect fundamental economic or social shifts. The agricultural training programme of the National Schools arose in the late 1830s and survived, largely unchanged, into the 1870s. Yet, in the same period, Irish agriculture underwent a substantial shift from tillage to grazing. Where populations were dense and holdings small, tillage survived, but the real growth was in cattle-raising on larger farms. Did instruction in the National School system adapt to that change? Even if it did, it is questionable whether improved methods would have made much impact upon farmers with small, usually rented, holdings of a few acres and large families to feed.

The serious dislocations caused by poor weather in the 1870s, combined with competition from cheap American grain, added insult to injury. In the end, Irish agriculture could only benefit from much larger holdings and more capital or, at least, cheaper labour. It possessed none of these things in the nineteenth century, thus cancelling much of the good that education might potentially have provided. This suggests that Irish agricultural education was designed for a fictitious population; it might have worked well in England or in parts of Scotland, but the economic and social realities of Ireland did not match the system of education. On the technical education side, what might have worked in Dublin, Belfast or Cork, at least in some industries, would not necessarily relate to the scale and style of industrial activity in smaller provincial centres.

It would be easy to blame the failure of agricultural and technical education upon those who directed it or who provided the instruction. Yet they were well-educated and well-intentioned men, whether they were local improvers such as the Rev. William Hickey and Dr John Hodges, or those on the national stage, like Sir Robert Kane. Of the latter group, the ubiquitous Dr William Kirby Sullivan is perhaps the archetype.[33] A Young Irelander in his youth, Sullivan went on to become a lecturer at the Museum of Irish Industry, professor at the Catholic University and at Glasnevin, freelance chemist, and, eventually, President of Queen's College, Cork, and a key figure in the founding of the Munster Institute. He was indefatigable in his efforts to improve the level of Irish education, agriculture and industry. Like his contemporaries, he believed that academic science is the royal road to industrial success. He was a paragon of the emergent Catholic middle class and, as such, probably out of touch with what the Irish farmer or industrial worker really needed. His intellectual vision differed little from that of contemporaries in many countries.

Where we find success stories like the Munster Institute or the Belfast Technical School, we also find both local leadership and more appropriate instruction. The top-down, state-directed efforts often appear to have been wasted efforts. This was just as true in Canada, where government adopted the National Schools model for agriculture and the Science and Art Department model for industry. Both failed to make much impact.

Let us note one final factor: general education. In Britain, many assumed – and continental experience seemed to show – that increased technical education would provide a stream of competent young people for industry, where their presence would enhance productivity and national competitiveness. This hope was to a large extent unfulfilled, because technical education was grafted upon a poorly organised and articulated educational structure (it was not even a 'system' on paper until 1902). Only a few progressive manufacturers, like Samuelson and Whitworth, welcomed such an invasion of talent into their shops. Ireland differed to the extent that it possessed an elementary education system, so that by the second half of the century, literacy was not a major obstacle for the Irish. Whilst many Irish manufacturers were as obtuse as their English fellows, there was so little Irish industry that no significant pool of jobs could have existed for potential graduates. Thus, technical, scientific or agricultural education was a stepping stone to other careers – teaching, politics, the civil service or, for the truly motivated, the liberal professions. All too often the graduates emigrated.

NOTES

1. The reader should consult the many excellent studies of Irish economic history, amongst them: L.M. Cullen, *An Economic History of Ireland since 1660* (London, 1972), especially chaps 5–6; Cormac Ó Gráda, *Ireland. A New Economic History 1780–1939* (Oxford, 1994); on the interplay of economic theory and policy, see R.D. Collinson Black, *Economic Thought and the Irish Question 1817–1870* (Cambridge, 1960).
2. A feel for the social climate early in the century may be obtained from R.B. McDowell, ed., *Social Life in Ireland 1800–45* (Cork, 1957).
3. See Simon Curran, 'The Society's Role in Agriculture since 1800', in J. Meenan and D. Clarke, eds., *RDS. The Royal Dublin Society 1731–1981* (Dublin, 1981). For background, consult Henry F. Berry, *A History of the Royal Dublin Society* (London, 1915).
4. An overview of Irish agricultural education is provided in an unsigned article, 'Outline of the History of Agricultural Education in Ireland to 1900', in Ireland, Department of Agriculture, *Journal* (1952–53), 3–27. A broader survey, by Austin M. O'Sullivan and Richard Jarrell, 'Agricultural Education in Ireland', is forthcoming in Norman McMillan, ed., *The Revolutionary Force in Irish Education* (Dublin, in press).
5. See Jonathan Bell and Mervyn Watson, *Irish Farming 1750–1900* (Edinburgh, 1986).
6. John Nicholas Murphy, *Ireland: Industrial, Political, and Social* (London, 1870), 410.
7. W.E. Vaughan, *Landlords and Tenants in Mid-Victorian Ireland* (Oxford, 1994).
8. A good summary is Sir Patrick Keenan, 'Agricultural Education', in William Coyne, ed., *Ireland, Industrial and Agricultural* (Dublin, 1902), p.137–45. For work of the National Board, see Matthew Burke, 'Agricultural Education in Ireland Under the National Board, 1831–70', (unpublished M. Ed. thesis, UCD, 1979). For a more general overview, consult Donald Harman Akenson, *The Irish Educational Experiment: The National System of Education in the Nineteenth Century* (London, 1970).
9. Ironically, the system's greatest impact was overseas, where many of its features were adopted and adapted in agricultural training in the USA, Canada and Australia. For example, the founding head of the Ontario Agricultural College was a graduate of the Glasnevin training school and its early programme was remarkably similar, even down to the diet for the students! See A.M. Ross, *The College on the Hill* (Guelph, 1974).
10. For the rebirth of agricultural interest, see Trevor West, *Horace Plunkett, Co-Operation and Politics* (Gerrards Cross, Bucks, 1986) and P. Bolger, *The Irish Cooperative Movement, its History and Development* (Dublin, 1971).
11. Anna Day, *More than One Egg in the Basket* (Cork, 1990), provides a history of the institution.
12. For background, see Gordon Herries Davies, *North from the Hook* (Dublin, 1995).
13. The syllabus-cum-examination scheme was taken over from the Society of Arts. Cole and Playfair were instrumental in its creation a decade earlier, from a plan of James Booth, a TCD graduate. See Harry Butterworth, 'The Science and Art Department Examinations: Origins and Achievements', in Roy MacLeod, ed., *Days of Judgement. Science, Examinations and the Organization of Knowledge in Late Victorian England* (Driffield, 1982), p.27–44.
14. There were, of course, larger political overtones. See Richard A. Jarrell, 'The Department of Science and Art and the Control of Irish Science', *Irish Historical Studies* 23:92 (Nov. 1983), 33–47.

15. See Garrett Scaife's article in this collection.
16. W.F. Barrett, *An Historical Sketch of the Royal College of Science from its Foundations to the Year 1900* (Dublin, 1907).
17. However, this was equally true for the Royal School of Mines in London. See Roy MacLeod, '"Instructed Men" and Mining Engineers: The Associates of the Royal School of Mines and British Imperial Science, 1851–1920', *Minerva* 32 (1995): pp. 422–39.
18. Department of Science and Art, *43rd Annual Report* (1896), vol. 30.
19. For an overview, see John Coolahan, *Irish Education: History and Structure* (Dublin, 1981), ch. 3.
20. The Department of Science and Art itself disappeared at the turn of the century, being absorbed into the new Board of Education. On the subsequent history of the Irish department, see D. Hoctor, *The Department's Story – A History of the Department of Agriculture* (Dublin, 1971).
21. See T.J. McElligott, *Secondary Education in Ireland 1870–1922* (Dublin, 1981).
22. Department of Science and Art, *43rd Annual Report* (1896), vol. 30.
23. Statistics are drawn from *Thom's Directory of Ireland* (Dublin, 1883).
24. The evidence taken in Ireland is in Vol. IV of the *Second Report of the Royal Commission on Technical Instruction*.
25. W.K. Sullivan, 'Outline of a Scheme for Technical Education for Ireland', in *Second Report of the Royal Commission on Technical Instruction*, Vol. III, p. cvii–cxix.
26. *Second Report of the Royal Commission*, Vol. I, p.538–9.
27. *Ibid.*, 503.
28. *Hansard*, 5 July 1887.
29. See Jim Cooke, 'Arnold F. Graves 1847–1930 Father of Irish Technical Education', *Education* 5:7 (1990): pp. 29–33.
30. By 1898, the total amount raised locally for technical education was about £5,600, to which was added just over £2,600 of Department of Science and Art funds. This contrasts with £843,000 expended in England and Wales at the same time. See Department of Science and Art, *46th Annual Report* (1899), p.lxviii.
31. *Ibid.*, 20.
32. *Ibid.*, 21.
33. On Sullivan, a useful biographical notice is T.S. Wheeler, 'Life and Work of William K. Sullivan', *Studies* 34 (1945): pp. 21–36; Wheeler also provides a portrait of Sullivan's comrade and sometime opponent, Sir Robert Kane: 'Sir Robert Kane – Life and Work', *ibid.*, 33 (1944): pp. 158–68, 316–30.

Chapter 8

Natural History in Modern Irish Culture

John Wilson Foster

THE NEGLECT OF NATURAL HISTORY

Histories of the development and practice in Ireland of science have barely begun to be written. In 1973, in the course of attempting an outline of such a history, Desmond Clarke wrote: 'The writer does not know of any history of science in Ireland'.[1] He meant a synoptic history: there have been histories of episodes in particular sciences (for example, by John Andrews in cartography and Gordon Herries Davies in geology)[2] – though most of those postdate Clarke – but Herries Davies' 1985 essay, 'Irish Thought in Science', was like Clarke's a pioneering outline, awaiting booklength amplification.[3] Perhaps before satisfactory synoptic histories of science can be written, there will have to be written extended histories of particular disciplines; either way, the history of science is still in its infancy in Ireland. Alan R. Eager's compendious *Guide to Irish Bibliographical Material* (London, 1980), and which lists sources as well as bibliographies, offers among its 9517 items only fourteen miscellaneous items under 'History of Science'.

In so far as science is international in methodology and in the dissemination of data and results, this historical lack is not wholly surprising. The local and social dimensions of such disciplines as mathematics, chemistry, microbiology, physics, biochemistry and astronomy are minimal, and the disciplines themselves prize currency and the 'advancement of learning' as their epistemological *raison d'être*. The 'backward look' is taken only for the retrieval of data or method that might be useful in a present project. Even so, narrative histories of the pure sciences as they have been introduced, developed, practised and taught in Ireland would be a distinct contribution to the island's self-reflection and our sense of its place through time in one vastly important sector of modern human experience.

119

Just as enlightening would be histories both of the application of the above sciences and of the applied sciences themselves, especially (if I may speak as a Northern Irishman) engineering, not in the incidental, specialised guises in which they fitfully exist already (histories of canals, railways, shipbuilding, the linen industry, and so on) but in broader, cultural forms, as though Robert Kane's once celebrated contemporary survey, *The Industrial Resources of Ireland* (Dublin, 1845), were given temporal as well as spatial form. Moreover, the recent explosion of computer technology, the building of transnational highways for information (its generation, proliferation, storage, deletion, reticulation and retrieval), threatens to render history, especially national or regional histories in narrative form, obsolete or cumbersome and yet behoves us to heighten our sense of the past in order to substantiate the present and reorientate us.

If computer technology with its simulative, even procreative capacities is in process of shifting our epistemological and even perceptual paradigms, science and industry have already done so for more than three centuries (most graphically in what we call 'revolutions' – the Scientific Revolution, the Industrial Revolution, the Communications Revolution). But the paradigm shifts have rarely been registered in modern Ireland even when they were happening. Such shifts always require not just their own history but a philosophy that can keep pace: nineteenth-century manufacture, for example, provoked outside Ireland – these are almost random examples – both the advocacy of Andrew Ure's *The Philosophy of Manufactures* (London, 1835) and the critique of Marx's *Capital* (1867), not to speak of Dickens's inspired but naive criticism in his novel *Hard Times* (1854), a novel which also takes a swipe at the contemporary and famous naturalist Richard Owen. We in Ireland have had the benefit of few histories and philosophies (and novels interested in science, applied or pure) and it is yet to be contemplated how far this deficit has augmented the already active forces of destabilisation on the island.

The more surprising fact is that even natural history (the systematic study of the earth and its productions: what is now inaccurately called natural science) has likewise suffered neglect in Ireland, not in practice (which has enjoyed at various periods a fairly high island-wide profile), but in reflection, in cultural reception and assimilation by assorted culture-givers. This neglect has been particularly unfortunate since Victorian times when field biology and evolutionary theory enjoyed great popularity. The technical biology practised today could not be expected, of course, to become part of popular or even high culture, save for blurred notions about DNA and genetic engineering. (These are in process of entering the popular consciousness, much as barely understood Freudian concepts and behaviourist concepts in psychology entered the popular consciousness earlier this century and as fuzzy ideas of computer capacities seem about to do the same.) Even the *stories* of highly advanced or technical sciences would be largely inaccessible to the lay person, explaining as they would have to do the complex workings (methods, theories, applications) of the sciences themselves and thereby resembling textbooks.

But one might have thought the case of natural history to be different. For one thing, nature study is a broadly popular pastime; for another, the natural world looms large in Irish literature, cultural iconography and the everyday experience of large numbers of people. Yet histories of the study of nature have not become part of the cultural curriculum of the island; Irish Studies does not yet claim them. Indeed, Irish 'society' and 'culture' are understood to comprise only the practice of politics, warfare, economics, literature and, occasionally, education. When customs and language are admitted, they are chiefly recruits to demonstrate or assail one definition or other of nationality (see, therefore, politics and warfare). The recent 3-volume *Field Day Anthology of Irish Writing* extends the received literary canon but only in the direction of (largely nationalist) political writing.[4] The anthology includes no nature writing. Giraldus Cambrensis (c.1146–1223), for example, is excerpted only because he wrote some unflattering things about the Irish in his *Topographia Hiberniae*, not because he wrote some fascinating early natural history of the island. William Thompson (1805-52), Robert Lloyd Praeger (1865–1953), William Spotswood Green (1847–1919) and Edward A. Armstrong (1900–78)[5] are nowhere to be found. Certainly nature writing is accessible, anywhere from novels through narrative survey reports and monographs to memoirs, but the editors would have had to indulge in some pioneering if elementary ransacking, for, outside of brief biographies of naturalists (for example, by Praeger), reminiscences (again, Praeger's are an example) and prefatory short histories (for example, Clive Hutchinson's account of Irish ornithology that introduces his *Birds in Ireland*), there are precious few histories of natural history upon which they could have laid their hands.[6]

CULTURAL SOURCES OF NEGLECT

It may be useful to ponder some possible reasons for the cultural neglect of science in Ireland. I offer them only as crude possibilities, for the investigation remains to be carried out; in several cases they are interpenetrating, and some concern only natural history.

Firstly, histories of science generally, even outside Ireland, have tended until recently to be regarded as luxury marginalia beside the text of scientific practice. And this may have been because science itself was seen in idealist terms, as Ophir and Shapin put it in 1991, with scientific ideas floating free, unlocalised, unsituated.[7] This seemed to be the chief way in which literature and science (the two great opposing, post-Enlightenment creative activities) differed. The history of writing was not just implicated in, but expressly re-enacted by, writing itself, which inscribes its own past even when departing from it: literature as referential, allusive, confluential, endlessly etymological. Whereas scientific practice depends absolutely on past, often immediately prior, practices (data collection, experimental results, hypotheses, theories), it is imagined when successful as cancelling the past; what is not wholly and essentially of the present, is by definition erroneous. The history of science might traditionally have been seen largely as a chronicle of error and asymptosis, an account of the past's approach to the present state of affairs, with the imminent future frankly even more interesting: projection and possibility more fascinating than history.

However, 'recognition of the situated character of our most highly valued forms of knowledge, and especially of science, has come into prominence over the past twenty years'.[8] It may be that this perceived situation has stimulated a new generation of science history outside Ireland. In which case, perhaps those in Ireland who are interested will belatedly respond to the stimulation, if only because for the Irish, the *situatedness* of knowledge (knowledge being power) is bite and sup. However, in the past, the recognition of the situatededness of scientific knowledge in Ireland has occasionally militated against science, for reasons explained below.

Secondly, if science until recently has been seen as essentially transnational, because unsituated, then it will have less appeal to a Euro-American historiography in which nationality and nationalism – especially since the mid-nineteenth century – are so important. That importance has been magnified in Ireland where nationalism has dominated a large portion of the island's intelligence and thoughtful discourse. To the shallow or exclusively political nationalist, science is of little use. As Clarke puts it, the scientist, 'unlike the writer, artist or patriot, is rarely a mere nationalist but always has been and always will be an internationalist'.[9] Although more thoughtful nationalists might derive some patriotic pride from Irish achievements in science – seeing Ireland in that field of endeavour as a nation among nations – thoughtful nationalists in Ireland have frequently been inclined to literature and the arts rather than to the sciences. Nationalism – though it fitted an international pattern established in various European movements: the French Revolution, English and German Romanticisms (as well as having its own internal dynamics) – necessarily emphasises what is distinctively Irish in the island's culture in order to justify political separatism: mentality, language, folklore, spirituality, literature, and survivals of all of these, and of a material culture before the 'Conquests'.

Where natural history is concerned (the subject matter, not the study), this can be no bad thing. In an essay for the *Nation*, the Young Irelander Thomas Davis (1814–45) challenges Irish travellers to postpone trips to the Continent in favour of travels through their own land. (Molly Ivors challenges Gabriel Conroy to do the same thing in one of the cruxes of James Joyce's great story, 'The Dead', written in 1907.[10]) In 'Irish Scenery', Davis might be seen as the father of modern Irish tourism, patron saint of Bord Failte.[11] But it is not just the picturesque and antiquarian riches he promotes; he remarks that 'The Entomology, Botany, and Geology of Ireland are not half explored' (not to speak of its ethnography, folk music, and social history). Although he reads here as the father also of the naturalists' field clubs (the first of which started up less than two decades later) – and the nationalist or at least regional/patriotic impulse behind these clubs remains to be investigated – Davis describes the landscape as *scenery*, a kind of cloak for the soul of Ireland. To travel amidst the scenery will instil love in the traveller and love of country will advance the nationalist cause. The scientific study of Ireland does not from his writings seem to appeal to Davis.

A history and analysis of relations in Ireland between nationalism and science would be worth undertaking. If nationalism inclines to ignore or reject science for being insufficiently rooted in place, it can also ignore or reject science when it is perceived as rooted in the wrong place and by the wrong people. For example, in another essay for the *Nation*, Davis proposes a shake-up and rearrangement of the literary, scientific and artistic institutions of Dublin. He suggests that there be only three: the Irish Academy, the Dublin Society (in each case he wishes the adjective of patronage, 'Royal', to be dropped: interestingly, it was dropped in neither case even when Ireland became a Republic in 1949) and a new Natural History Society or Academy (with a 'specially Irish museum'). He particularly wishes the (Royal) Irish Academy to 'divorce' itself from its scientific department, retaining its interest only in antiquities and literature (which of course can be seen as ornaments of Irishness). He would 'require' Trinity College (the college of the anti-nationalist Ascendancy) 'to form some voluntary organisation for the purpose'. Science, it would seem, is weighed in the balance by a nationalist and found wanting: best return it to its proper enclave.[12]

Thirdly, Irish nationalism has required due recognition of, ethnically and culturally speaking, the Catholic (aka 'native') people of the island. Although before the secession of twenty-six counties from the United Kingdom there were practitioners of science in that part of Ireland who were from the Catholic portion of the population – and even some significant achievers, such as Bernard O'Connor (1666–98) in medicine and Nicholas Callan (1799–1864) in physics – it may have been the case that the profile of science in native Ireland was not high enough to warrant the adoption of Irish science into the nationalist perspective.

Indeed, science was perceived as an activity of that part of the Irish population that was on the whole anti-nationalist. As Richard Jarrell remarks: 'there were two Irelands, and almost all the scientists were Anglo-Irish. Trinity College, the Royal Dublin Society, and the Royal Irish Academy were near exclusive preserves of the minority . . . In centres outside Dublin, especially in Cork and Belfast, the scientific leaders were equally

Anglo-Irish. Were there scientists in the "other" Ireland? There were, of course, but few in number'.[13] Despite the historical sequence of Anglo-Irish leaders of nationalism, the latter regards itself as rooted in a certain ethnicity, a certain religion and certain social classes, beyond the historical exercise of mere political power. Science itself was in danger of being ethnicised as foreign (not truly Irish), categorised as anti-religious (not acceptably Catholic) and classified as socially distant. The Anglo-Irish had their 'places of knowledge' (to borrow Ophir and Shapin again) – estates, gentlemen's houses, observatories, botanical gardens, rooms of curiosities – and so what Ophir and Shapin say would seem vividly to apply to the Irish case: 'the place of knowledge is implicated in the network of relations between knowledge and power, in the distribution of knowledge in society, in perceptions of its validity and legitimacy'.[14]

After their political and then numerical decline once the Irish Free State and then the Republic of Ireland came into being, the Anglo-Irish seem to have become embarrassed about their achievements, including their scientific achievements. Yeats was a notable and eloquent exception in this matter, though he had little time for scientific achievements: see for example, his memorably defiant speech to the Senate of Saorstát Éireann in 1925 occasioned by the bill to outlaw divorce – the celebrated 'We are no petty people' speech.[15] He had only himself to blame for the circumstances that necessitated his speech. For he and the other literary Anglo-Irish, along with the more favourably disposed nationalists, conspired in the years leading up to the founding of the Irish Free State to invent a cultural phenomenon we might call the Anglo-Irish Twilight. This romantic creation – an Ascendancy class sinking into poetic oblivion – retroactively obscured the achievements of the Anglo-Irish fashioned during the daylight hours of reason. After independence, histories of those achievements would have seemed imprudent. For their part, many of those belonging to the majority population were deeply ambivalent about science (at once admiring, embarrassed and hostile) because of their understandable confusion of it with political power of a certain kind. Dropping out of historical self-reflection, science dropped out of the received culture of a society that regarded itself as having supplanted the power of the Anglo-Irish.

A word about social class that complicates the picture of two Irelands and suggests at least a third. When we speak of the Anglo-Irish, we are speaking largely of the gentry and upper middle class (minor nobility, great landowners, and merchants and professionals of high social standing), but in Belfast and environs in the nineteenth century, we would speak less of the Anglo-Irish than of the Protestant middle and lower middle class. As Jarrell reminds us, 'the middle class, however narrowly and broadly it is defined, is the class of industrialisation and science',[16] and it is perhaps the case that achievements in engineering and the applied sciences are even less likely to feature in the kind of historiography (still with us) that is a descendant of polite literature, particularly when they were associated with a culture or subculture hostile to the newly empowered majority culture in the south of the island and increasingly disowned by the Anglo-Irish in the south.

Fourthly, when we speak of the Anglo-Irish, we are also speaking of a *Protestant* population on the island of Ireland. The historical European connections between science and Protestantism that obtained until recently (of which the systematic study of Nature that Natural Theology sponsored is but one episode) surely reinforced the idea that science was rather alien to the 'real' Irish because practised by the Anglo-Irish and therefore, at least in terms of cultural history if not in narrow practice, to be given a wide berth.

Protestantism of course has had its own quarrel with science, including biology, for example Creationism's quarrel with Darwinism. It is retrospectively appropriate that the aggression of evolutionists towards Christianity should reach a climax in 1874, in Belfast, an important centre of Protestant theology (not just Anglican but even more importantly Presbyterian), with Carlow-born John Tyndall's presidential address to the British Association for the Advancement of Science (the so-called 'Belfast Address') which Tess Cosslett judges 'more important than the Huxley/Wilberforce clash in 1860'.[17] The equally aggressive (and often impressively intelligent) response by local divines – some of them amateur scientists – and the long-standing prevalence of Protestant missionary zeal in Ulster almost certainly had its inhibiting effect on those who would have incorporated post-Darwinian, secular, even heretical natural history into the Irish cultural scheme of things.[18]

But if the effect of doctrinal Protestantism on science-as-culture was in Ulster prophylactic (coincidentally, the year of 1859 saw both the publication of *The Origin of Species* and the great evangelical Revival that erupted in the north of Ireland and spread to Britain), Protestantism in its broader cultural form (sponsoring the Enlightenment, rationalism and secular humanism) not only accommodated science but welcomed it, indeed gave birth to much of what we call modern science. Although often thought of as a cooperative project, science has frequently depended upon, and been an expression of, high individuality of a kind that grew out of both Protestantism and the Enlightenment and was opposed by the counter-Enlightenment.[19]

Natural Theology was the branch of Protestantism that sponsored scientific observation; its key texts included Thomas Burnet's *Sacred Theory of the Earth* (1681–89), John Ray's *The Wisdom of God as Manifested in the Works of the Creation* (1691), William Derham's *Physico-Theology* (1713), Cotton Mather's *The Christian Philosopher* (1721) and William Paley's *Natural Theology* (1802). Evolution of course damaged the collaboration. Lyell, Darwin, Wallace, Tyndall, Haeckel and Huxley brought the chronic rivalry between science and religion (however much it was denied by figures as respected as Richard Owen and Philip Henry Gosse) into the light of common day where it could no longer be ignored. True, henceforward culture in Britain and Protestant Ireland was a divided empire. Indeed, it was divided three ways after English literature too broke with science, as well as with religion, under the tutelage of Matthew Arnold, a Victorian sage whose influence rivalled that of the scientists above. There is no better, or better written, popular account of the tensions and crises involved in this triangular contest in Victorian Britain than the literary critic Edmund Gosse's memoir of his father, the naturalist and Plymouth Brethren minister P.H. Gosse (*Father and Son*, London, 1907). But in Britain and Protestant Ulster, science and its theological enemy at least had a mutual cultural origin, and even if the *history* of science did not gain a prominent place in the cultural curriculum, the *practice* of science became part of the cultural mainstream whilst the *theory* of organic evolution became one of its important tributaries.

Fifthly, Catholicism in Ireland composes a social as well as religious complex significant enough to have greatly affected the history of science (both its practice and its story) in the island. This is a touchy subject, since in Ireland only practising or lapsed Catholics have a licence to criticise the Catholic church without being labelled sectarian. Yet an historian of culture must not be squeamish in asserting that the Catholic church in Ireland has not on the whole encouraged science or explicitly entertained scientific explanations of cosmic mechanisms and the evolution of life on earth. That church has been a counter-Enlightenment force and has generally obstructed the introduction and development of Enlightenment values in Ireland long after they had become part of the common intellectual currency of Protestant Europe and America.

The hostility is rooted in the particular theology of that church but also in the power the church exerted in Ireland over social relations and the exercise of published thought. Examples of that hostility would constitute a social history in themselves. We could cite an example from the historical case I have already adduced with reference to Protestantism in Ulster. According to Cosslett, Tyndall in 1874 had 'a particular clerical target in mind, connected with the venue of the meeting in Belfast: the Irish Catholic hierarchy had just thrown out a plan to include physical science in the curriculum of the Catholic University'.[20] But of course, any examples are merely part of a larger contest in Catholic Ireland (as in other Catholic countries with a large historical peasantry) between the church and science (and technology, especially medical technology) in their ethical, epistemological and social guises.

Against this picture of an anti-scientific and powerful church can be placed Clarke's claim that although Maynooth College, Catholic Ireland's venerable chief seminary, is thought of as primarily a theological institution, 'it has had a strong scientific tradition in what used to be called applied science'.[21] And Jarrell records that he has 'been unable to detect any significant antiscience bias on the part of the clergy in . . . Ireland during the nineteenth century'.[22] The history of the relations in Ireland between science and Roman Catholicism (the church and laity alike) is pretty much a fallow field but I expect that the new circumstances surrounding the Catholic church in Ireland will soon cause this field to be ploughed, sowed and reaped. When it is, the harvest should be fascinating for the Irish cultural historian.

A sixth impediment to an Irish scientific culture came from an unlikely source – literary Protestants – and at an unlikely time – a generation after the Victorian evolutionists had successfully contested the cultural and intellectual hegemony of Christianity. I am speaking of the great movement of cultural nationalism that was the Irish Revival led by Standish James O'Grady, Douglas Hyde, AE, W.B. Yeats and others (the chief figures being Protestant), a movement which began around 1880 and lasted, surely not coincidentally, until the decade that saw political independence in twenty-six of the thirty-two counties of the island. This movement was a belated Irish expression of European folk nationalism (which began in Germany in the 1770s), European literary Romanticism (that flourished in England between roughly 1780 and 1830) and – as Easter 1916 showed – European political revolution (which in its modern, nineteenth-century form began with the French Revolution of 1789).

The presiding genius after O'Grady faded in importance, and a figure who has had an immense general cultural impact on twentieth-century Ireland (from poetry to folk culture, from mysticism to coinage, from heroism to tourism), was Yeats, for whom a counter-Enlightenment and the overthrow of science were enduring articles of faith since youth. As a very young man he had been a keen naturalist (especially enthusiastic about lepidoptera) but early discovered the works of the mystics and, as he later explained, turned his back on Haeckel, Darwin, Tyndall and Huxley, soon after converting to Celticism and Irish cultural nationalism.[23] His accumulating cosmic philosophy, *A Vision* (the first version of which was published in 1926), was a Theory of Everything that sought to reduce the Enlightenment and science to mere episodes in a spiral or gyratory cosmic motion, and not particularly salutary episodes either; like Nietzsche, whose philosophy encouraged his own after 1900, he appeared to welcome the prospect of a cleansing violent reversal of humanism and rationalism.

To the extent that science was international instead of national, rationalist instead of spiritual, empirical instead of mystical, it could hardly be part of a revival headed by Yeats, O'Grady, Hyde and AE (though all of these men had their practical, organisational side). Here was a sixth antiscientific factor in Irish culture that after a century is only now beginning to wane.

A rare attempt to marry science to Celtic nationalism, and one it is almost bathetic to mention in the same breath as Yeats's magisterial works, was Michael Moloney's curious book, *Irish Ethno-Botany and the Evolution of Medicine of Ireland*. It essays a history of a national (or indigenous) medicine, claims that the study of 'Celtic nature creeds' should enter the economy and the educational curriculum, that the centre of Irish medicine should be ethno-botany, and that 'a study of . . . native flora *in his own tongue* [i.e. the Irish language] will enable the student to inherit some of the scientific, literary, aesthetic, and religious possessions of the race'.[24] Long before Foucault, Moloney believed that science was 'situated' and 'implicated in the network of relations between knowledge and power', to quote Ophir and Shapin again. Despite Arnold and before C.P. Snow, Moloney saw no essential division between science and literature, and it was the nationalist project that enabled him to do so. This contribution to Irish nationalism, however, seems to have petered out: *Irish Ethno-Botany* is a peculiar and rare book and does not appear to have been emulated in other fields of natural history. Under the circumstances, one might almost enter a note of regret that this has been the case.

SCIENCE AND MODERN IRISH CULTURE

Given its cultural and not merely literary eminence, I would like now to say something about twentieth-century Irish literature and its relations with science. For the most part, Irish writers since the Revival have been antiscientific, and writing until recently has dominated Irish culture and the image of that culture abroad. (Irish film and Irish folk rock music have begun to challenge it.) And until recently, too, Irish literature derived chiefly from Romanticism via the Irish Revival. Raymond Williams in *Culture and Society* has identified literary antiscientism from Romanticism onwards.[25] (This despite the interest of the Shelleys in electro-magnetism and galvanism and the importance of Humphry Davy to the Shelleys as well as to Coleridge.) The Romantic coloration of Irish expressive culture has been, as I have implied, an inhibiting factor in the habilitation of science and its history in Irish culture.

Of course, one legacy of Romanticism (and of the vestigial pagan and medieval Christian faiths of the island) has been the writers' frequent, almost reflex re-creation and celebration of Nature for its healing, aesthetic and spiritual properties. There is also the rich possibilities of Nature's symbolism, everything from James Joyce's use of a snow-covered Galway to suggest paralysis, death and vision in 'The Dead' through Frank O'Connor's use of bogland as a vehicle for violent indignity in his powerful story, 'Guests of the Nation', to the imagery in many writers of neglected gardens and demesnes to suggest the decline of the Ascendancy.[26] And there is, of course, Nature's general usefulness as background, as visual exposition (what is called in cinematography 'establishment shots') and as closure at the ends of scenes, chapters or whole narratives. But this literary exploitation of Nature has on the whole involved the repudiation of a scientific approach.

Now it is arguable that the scientific approach has nevertheless exerted an influence, even in unlikely circumstances. Despite his fierce antiscientism, Yeats in the 1880s and 1890s classified fairies and fairy tales in an almost Linnean fashion (*Fairy and Folk Tales of the Irish Peasantry* (1888) and *Irish Fairy Tales* (1892)) and even lectured to the Belfast Naturalists' Field Club on the latter subject, during which he made claim for the 'scientific utility' of the study of fairy belief.[27] I have argued recently that Oscar Wilde – the aesthete inheritor of the Symbolist strain of Romanticism – transposed his early Oxford enthusiasm for contemporary science, especially biology, into a dominant concept of form that emulated science in its attempted precision and nicety of nomenclature.[28] Joyce, too – an early naturalist in the literary sense, but also a Symbolist of sorts – read across the two cultures and his scholasticism frequently issues in *Ulysses* (1922) as mock-science. I am thinking, for example, of the allegorical role of gynaecology in the Rotunda Hospital episode.[29] In *A Portrait of the Artist as a Young Man* (1916), Joyce's hero Stephen Dedalus rejects Darwinism as an explanation of aesthetic attraction.[30] Nonetheless, Joyce's aesthetic – like Wilde's – depends upon the very science it supplants. (Joyce, though, disliked the countryside as much as did Wilde, and could not countenance natural history. In *A Portrait*, it is an unprepossessing student with a 'pallid bloated face' who is taking botany at Stephen Dedalus' university and is a member of the field club, mainly it seems because it allows him respectably to go on outings with women students.[31]) And of course Shaw was abreast of evolutionism: among other plays, there are *Back to Methuselah: A Metabiological Pentateuch* (1921) and *John Bull's Other Island* (1904), in which a reference to 'the late Professor Tyndall' occurs.[32]

Moreover, some Irish writers have drawn upon scientific sources. Brian Friel's use of John Andrews' historical cartography in *A Paper Landscape: the Ordnance Survey in Nineteenth Century Ireland* (Oxford, 1975) for his play *Translations* (1981) is one example. Some writers have had an early interest in natural history and its organised study; Sean O'Casey is one such. Some have fashioned analogies or parallels to scientific method in their work; Yeats's *A Vision* is an attempt to substitute his own cosmology for that generated between Galileo and Einstein.

Still, the tendency of Irish writing since Davis, Mangan, Ferguson (to choose the nineteenth-century trio whose tradition of romantic literary nationalism Yeats said he was following) has been anti-scientific, and that writing has been more psychoculturally important in Ireland than its counterpart in England. It is arguable, for instance, that the profitable imagery of Ireland generated by Bord Failte and the Northern Ireland Tourist Board, although deriving originally from the eighteenth-century Topographical (or Prospect) poets and cultists of the Picturesque, whose enthusiasms were then passed through Davis's nationalist filter, was brought to copious bloom by Yeats, John Millington Synge, and the Revival. In Synge, incidentally, we have the Romantic translation of

Nature into a lush celebrative rhetoric but also – in *The Playboy of the Western World* (1907) – a register of the Irish country-dweller's indifference to nature except in practical terms: the eponymous hero Christy Mahon's interest in 'felts and finches' is evidence of his contemptible simpleness and effeminacy.[33] ('Felts' was a colloquial Irish word for thrushes.)

Writers who have been sympathetic to science or who turned it directly and positively to account, have been minor writers outside the Romantic mainstream. For example, the Rev. William Hamilton Drummond in *The Giant's Causeway: A Poem* (1811) wrote a long, quasi-epic work in the Topographical tradition and among other things it learnedly versifies the conflicting theories of Neptunism and Vulcanism. Earlier Topographical poets drew on scientific surveying in the arrangement of their poems.[34] Or they were those writing before the Enlightenment period or before Romantic nationalism. Such figures were either writers in the orthodox sense (Oliver Goldsmith, for example, was the author of *An History of the Earth and Animated Nature* in 1774) or men of other pursuits and on the fringe of literature: political figures or scientists (William Molyneux in the seventeenth century was both) or antiquarians (from Archbishop Ussher in the late sixteenth and early seventeenth centuries to Eugene O'Curry in the nineteenth). One detects little animus against science or method in the writers D'Arcy McGee discusses colourfully in *The Irish Writers of the Seventeenth Century of 1846*.[35]

The Healthy Fact of Science

However, things might be changing. It is true that Yeats's choice in his poem 'Sailing to Byzantium' between the salmon-falls of Ireland (i.e., an unexamined life 'in' nature) and a sixth-century city of the imagination (i.e., an examined life 'out of nature') is replayed in 'Death of a Naturalist' (a manifesto poem in Seamus Heaney's first collection of that name, published in 1966) as a choice between a mythopoeic relationship to nature and a scientific study of nature. But there has recently been some adverse reaction to the confident simplicity of Heaney's choice.[36] Irish Nature as a spiritual-aesthetic-female-nationalist state of affairs – an immense ethnocentric convenient shorthand – is no longer a supportable idea. Nor is science seen any more as essentially or tendentially anti-Irish or even anti-nationalist. The Irish novelist John Banville, fictional biographer of Copernicus, Kepler and Newton, is looking more abreast of the *Zeitgeist*, his work perhaps a contributory sign of it.[37]

Certainly up until recently, what Christie and Shuttleworth said in 1989 remains broadly true, especially in Ireland: 'Our culture and criticism tends to pair literature and science as opposites'.[38] In the Anglo-American cultural complex (that incorporates much of Ireland) this derives chiefly from Arnold. Arnold championed the centrality of classical and modern literature in the university curriculum and in culture, against the emerging claims of science; in his essay 'Literature and Science' (1885), he answered the criticisms levelled at his position by T.H. Huxley in his own essay, 'Science and Culture' (1882).[39] The gap between Arnold and Huxley succeeded this century to that between Leavis and Snow, and it was Snow not Leavis who regretted the duality of culture.[40]

Similarly, it has largely been the Irish naturalists who have attempted to cross the great divide. Curiously, three Irish naturalists were also Shakespeareans: Robert Patterson (1802–1872) wrote *Letters on the Natural History of the Insects Mentioned in Shakespeare's Plays* (1838) and his own poems were anthologised; Henry Chichester Hart (1847–1908), expeditionary botanist, was an editor of the prestigious Arden series of Shakespeare editions; Edward A. Armstrong the ornithologist (1900–1978) in his book *Shakespeare's Imagination* (1963) pioneered the analysis of cluster imagery in Shakespeare's plays. William Thompson (1805–1852), perhaps Ireland's greatest all-round naturalist, was President of the Fine Arts Society in Belfast and patronised the poet Francis Davis, one of the nineteenth-century 'rhyming weavers'. J.J. Murphy (1827–1894), the Christian evolutionist, author of *Habit and Intelligence* (1869) and *The Scientific Bases of Faith* (1873) and other works, was President of Belfast's Literary Society and published *Sonnets and Other Poems* in 1890. Thomas Corry, drowned as a young man like a botanical Shelley, published with S.A. Stewart the still standard (if much revised) *Flora of the North-east of Ireland* (1888) and was the author of a volume of poetry, *A Wreath of Wind-flowers* (1882).

This apparently healthy obliviousness to the 'Two Cultures' would of course have had an element of facilitating philistinism in it, and Belfast's nineteenth-century inter-disciplinary culture would have had in it an element of Dickensian Gradgrindery. Still, the history of intellectual Belfast – continuous with intellectual cities in Britain, especially with those in the industrialised midlands and north of England and lowlands of Scotland (as the work of David Allen demonstrates)[41] – has been driven underground, and any history of science in Ireland has to bring it to the surface. It strengthens any claim that the *fact* of science in Ireland is healthier than the prevailing *fiction* of its sparsity. For researchers in Belfast, William Gray's scarce book, *Science and Art in Belfast* (1904), is the essential starting-point.

Meanwhile, the miscellaneous social reasons Michael Viney recently adduced for the neglect of natural history in modern Ireland apply mainly to the south of Ireland: the paucity of towns; the absence of industrialism; general scission from older benign Anglo-Irish views of Nature; the collapse of faith in Nature after the catastrophe of the Famine; the small percentage of Dissenters, Quakers and evangelicals in the population.[42] What Viney calls the Irish View of Nature – our seventh adverse factor – is not conducive, he holds, to its systematic study. But neither part of Ireland can afford to neglect the story of that study as it has been conducted since the early naturalists, or to applaud the past cultural deficit when such study was discouraged or ignored by the culture-givers on the island. The critical threat to the natural environment alone makes the case. But cultural diversity and the desirability of cultural integrity make the case equally.

NOTES

1. Desmond Clarke, 'An Outline of the History of Science in Ireland', *Studies* 62 (1973): pp. 287–302, p.287. Since then, there has been Gordon Herries Davies (see below) and a pioneering but slight book by John R. Nudds et al., *Science in Ireland 1800–1930* (Dublin, 1988).
2. See J.H. Andrews, A Paper Landscape: *The Ordnance Survey in Nineteenth-Century Ireland* (Oxford, 1975) and G.L. Herries Davies, *Sheets of Many Colours: The Mapping of Ireland's Rocks 1750–1890* (Dublin, 1983). The latter has now been supplemented; see the same author's *North from the Hook: 150 Years of the Geological Survey of Ireland* (Dublin, 1995).
3. Gordon L. Herries Davies, 'Irish Thought in Science', in Richard Kearney, ed., *The Irish Mind: Exploring Intellectual Traditions* (Dublin, 1985), pp. 294–310.
4. Seamus Deane, ed., *Field Day Anthology of Irish Writing* (Derry, 1991).
5. The American zoologist William Beebe wrote that Armstrong's *Birds of the Grey Wind* (1940) 'puts the author well in the forefront of natural historian *belles-lettres*' and that his *Bird Display* (1942) 'assures him no uncertain position among the company of scientific ornithologists,' see Beebe, *The Book of Naturalists: An Anthology of the Best Natural History* (Princeton, 1988).
6. See Robert Lloyd Praeger, *The Way That I Went* (Dublin, 1937); Praeger, *Some Irish Naturalists* (Dundalk, 1949); C.D. Hutchinson, *Birds in Ireland* (Calton, 1989).
7. Adi Ophir and Steven Shapin, 'The Place of Knowledge: A Methodological Survey', *Science in Context* 4 (1991): pp. 3–21, p. 3.
8. *Ibid.*, p. 4.
9. Clarke, 'An Outline of the History of Science in Ireland', p. 287.
10. 'The Dead' is the last story in Joyce's collection *Dubliners* (London, 1914).
11. Thomas Davis, 'Irish Scenery' in *Prose Writings: Essays on Ireland*, (London, 1889), pp. 184–7, see p. 184 and p. 186.
12. Thomas Davis, 'Institutions of Dublin', *ibid.*, pp. 166–73, see p. 172.
13. Richard Jarrell, 'Differential National Development and Science in the Nineteenth Century: The Problems of Quebec and Ireland', in Nathan Reingold and Marc Rothenberg, eds., *Scientific Colonialism: A Cross-Cultural Comparison* (Washington, D.C., 1987), pp. 323–350, see p. 341–342.
14. Ophir and Shapin, 'The Place of Knowledge', p.15, referring to Yaron Ezrahi, *The Descent of Icarus: Science and the Transformation of Contemporary Democracy* (Cambridge, Mass., 1990).
15. Reprinted in *Yeats' Senate Speeches*, ed. Donald R. Pearce (Bloomington, Indiana, 1960), pp.89–102.
16. Jarrell, 'Differential National Development', p.325–6.
17. John Tyndall, *Address Delivered before the British Association Assembled at Belfast* (London, 1874). Tess Cosslett, introduction to Cosslett, ed., *Science and Religion in the Nineteenth Century* (Cambridge, 1984), p.9.
18. For a miscellany of Presbyterian responses to evolution, see *Science and Revelation: A Series of Lectures in Reply to the Theories of Tyndall, Huxley, Darwin, Spencer etc.* (Belfast, 1875).
19. See Dorinda Outram, 'Heavenly Bodies and Logical Minds', *Graph*, (Spring 1988): pp. 9–11.
20. Cosslett, *Science and Religion in the Nineteenth Century*, p.173.

21. Clarke, 'An Outline of the History of Science in Ireland', p.298.
22. Jarrell, 'Differential National Development', p.341 (but see Cosslett, *Science and Religion in the Nineteenth Century*).
23. W.B. Yeats, 'Four Years: 1887–1891', in *W.B. Yeats: Selected Prose*, ed. A. Norman Jeffares (London, 1976), p. 50.
24. Michael Molony, *Irish Ethno-Botany and the Evolution of Medicine in Ireland* (Dublin, 1919), p. 9.
25. Raymond Williams, *Culture and Society, 1780–1950* (London, 1958).
26. 'Guests of the Nation' is the title story of an O'Connor collection (London, 1931).
27. Yeats's lecture to the BNFC was given on November 20, 1893 and was reported in *Proceedings of the Belfast Naturalists' Field Club* (1893–94): pp. 46–48.
28. John Wilson Foster, 'Against Nature? Science and Oscar Wilde', *University of Toronto Quarterly* 63.2 (1993–4): pp. 328–46.
29. *Ulysses* (Harmondsworth, 1971), pp. 380–425.
30. *A Portrait of the Artist as a Young Man* (Harmondsworth, 1960), p. 208.
31. *Ibid.*, p. 210.
32. *John Bull's Other Island and Major Barbara* (London, 1907), p.38.
33. *The Playboy of the Western World* in John M. Synge, *The Complete Plays* (New York, 1960), p. 49.
34. I discuss Drummond and other Irish Topographical poets in 'The Topographical Tradition in Anglo-Irish Poetry', *Irish University Review* 4.2 (1974): pp. 169–187, and the connection between scientific surveying and English Topographical poetry in 'The Measure of Paradise: Topography in Eighteenth-Century Poetry', *Eighteenth-Century Studies* 9.2 (1975/76): pp. 232–56.
35. Thomas D'Arcy McGee, *Gallery of Irish Writers: The Irish Writers of the Seventeeth Century* (Dublin, 1846).
36. Sean Lysaght, 'Heaney vs Praeger: Contrasting Natures', *The Irish Review* 7 (1989): pp. 68–74; see Seamus Heaney, *Death of a Naturalist* (London, 1966).
37. John Banville, *Doctor Copernicus* (London, 1976); *Kepler* (London, 1981) and *The Newton Letter* (London, 1982).
38. John Christie and Sally Shuttleworth, *Nature Transfigured: Science and Literature, 1700–1900* (Manchester, 1989), p.1.
39. T.H. Huxley, 'Science and Culture', reprinted in Huxley, *Science and Education: Essays* (London, 1893), pp. 134–159; answered by Matthew Arnold, 'Literature and Science', in *Discourses in America* (reprinted London, 1896), pp. 72–137.
40. C.P. Snow, *The Two Cultures and the Scientific Revolution* (London, 1959), answered by F.R. Leavis, *Two Cultures? The Significance of C.P. Snow* (London, 1962).
41. David E. Allen, *The Naturalist in Britain: A Social History* (London, 1976)
42. Michael Viney, 'Woodcock for a Farthing: the Irish Experience of Nature', *The Irish Review* 1 (1986): pp. 58–64.

Chapter 9

Twilight and Dawn for Geography in Ireland[1]

Anne Buttimer

INTRODUCTION

Toward the close of the nineteenth century Pitirim Kropotkin wrote:

> Geography . . . must teach us, from our earliest childhood, that we are all brethren, whatever our nationality. In our time of wars, of national self-conceit, of national jealousies and hatreds ably nourished by people who pursue their own egotistic, personal, or class interests, geography must be – in so far as the school may do anything to counterbalance hostile influences – a means of dissipating these prejudices and of creating other feelings more worthy of humanity. It must show that each nationality brings its own precious building-stone for the general development of the commonwealth, and that only small parties of each nation are interested in maintaining national hatreds and jealousies. [2]

These idealistic sentiments were voiced at a time when geography, for a variety of political and commercial reasons, was claiming significant attention in academy, officialdom and popular press. Imperial geographies projected images of people and places beyond the pale of the 'home' civilisation as Others rather than brothers. The connections between geography and empire are well-worn themes in the history of the discipline: in the bolstering of national identity and socialisation of youth, the training of personnel for military, mercantile or missionary expeditions, and the shaping of world images, geography played an ideologically important role. How newly-independent states or changed regimes within a state have set about the re-fashioning of their geography texts makes an equally interesting tale. [3]

This paper examines elements of geographical thought[4] involved in debates about education and national development during that 'twilight' period preceding and following independence in the Republic of Ireland

(1922). For an island so renowned for its natural beauty and historically acclaimed literary traditions, it seems quite amazing that geography as an academic discipline had such a late start. Gordon Herries Davies cited an *Irish Times* editorial (13 January 1923) asserting that the only European countries where geography was not accorded university status were Albania, Greece, Latvia, and Ireland.[5] Why did it take so long to convince Irish universities, and the National University in particular, about the value of geography as academic discipline? As in other post-colonial settings, the difficulties were not primarily of an epistemological nature; rather they illustrate the salience of gatekeepers and the vicissitudes of political history in the negotiation of ideas and context.[6]

Two main themes will be explored. First, on the level of intellectual style, a critical distinction is made between the integrated world views which fostered political movements prior to independence, versus the dispersed world views which fostered sensitivity to landscape and culture and the affairs of state afterwards. I suggest that the integrated world views of mechanism and organicism were associated with broader horizons of concern, whereas the more dispersed views of formism and contextualism were associated with more circumscribed horizons.[7] Secondly, exploring the level of popular appeal a distinction is drawn between the rhetorics of 'liberation from' versus 'freedom to': emphasis on the former often serving to create structures which still 'wore the mantle of the oppressor'. Given the dramatically changing political contexts for the practice of geography in the late twentieth century, there may be lessons to be drawn from these stories.

ORIGINS OF MODERN GEOGRAPHY IN IRELAND

Geography was taught in Irish monastic schools from early medieval times and was indeed a popular subject in the hedge schools during the period of penal laws.[8] The origins of geography as professional field in modern times has been traced to 1731 when a Royal Society was founded 'for Improving Husbandry, Manufactures, and Other Useful Arts and Sciences' (after 1820 known as The Royal Dublin Society), the first Society of its kind in Europe.[9] Members showed a keen interest in the Irish landscape, its resources, traditions, and potential usefulness within the Commonwealth. Statistical surveys of Irish counties were conducted and by 1830 a survey of land quality for the island was completed by Richard Griffith et al.[10] And from the eighteenth century on, philosophical and historical societies in Dublin engaged in pioneering research in geological, botanical and cultural features of the Irish landscape.

The nineteenth century also witnessed some steps toward the foundation of geography and geology in Ireland. Railroads were to be built and canals dug; survey and mapping, e. g. , by the Ordnance Survey, were encouraged. In 1845 the Geological Survey of Ireland was inaugurated

Twilight and Dawn for Geography in Ireland

and a 1" to the mile geological map of the island was complete in 1890. [11] It was in 1845 also that the Dublin Museum of Economic Geology was founded. Later to be named as Museum of Irish Industry (1847), College of Science for Ireland (1865), and finally absorbed within University College Dublin (1926), this was the first institution to establish a chair of geology in 1890. The Queens' Colleges of Belfast, Cork, and Galway were founded in 1845 and physical geography was taught there in the context of geology departments. At Trinity College Dublin, geology acquired departmental status in the context of the engineering school in 1841.

The turn of the twentieth century witnessed the establishment of geography as university discipline in most European countries. After 1871, International Geographical Congresses continued to provide a quadrennial forum for the exchange of research ideas. During the first quarter of the twentieth century Halford Mackinder secured a firm foundation for geography at University and school levels in England. Vidal de la Blache fostered *la géographie humaine* on the twin foundations of geology and history in France, delivering regional accounts of life and landscape which to this day command applause. Throughout Scandinavia and peripheral regions of Europe, geography flourished not only through exploration and geomorphology, but also as *Heimatkunde*, knowledge of home areas, local studies geared toward fostering the sense of patriotism and nationhood. By 1922 when the International Geographical Union was formally established and geography was taught in 120 European universities, there was still no department of geography in Ireland.

FOUNDERS AND GATEKEEPERS

The first decades of the twentieth century were indeed a 'twilight' time, politically and intellectually in that land. Focus on three particular voices seeking to argue the case for geography during that time – those of Grenville Arthur James Cole (1859–1924), Horace Plunkett (1854–1932), and Timothy Joseph Corcoran (1872–1943) illustrate the importance of context and the perennial puzzle of 'gatekeepers' in the negotiation of disciplinary orthodoxy.

GRENVILLE ARTHUR JAMES COLE (1859–1924)

Grenville Arthur James Cole, graduate of the City of London School, and for twelve years demonstrator at the Royal School of Mines, was appointed as Director of the Geological Survey of Ireland in 1905. [12] Cole was Ireland's first professional geographer. He travelled widely, mostly on bicycle, throughout the length and breadth of the island, and across Europe. Already in 1895 he published *Open-air Studies: An Introduction to Geology out of Doors*, and a 19-page pamphlet on *Scenery and Geology in Country Antrim*. As researcher he had conducted comparative studies of glacial landforms in Spitzbergen and Ireland, and proposed a novel theory on the origin of the Liffey. His international network was impressive: he invited his German colleague, Albrecht Penck, to become life member of Royal Irish Academy in 1906, and in 1911 accompanied an American geographer, William Morris Davis, through S. W. Ireland on the first steps of the trans-European excursion organised by the International Geographical Union. [13]

Geography was an integral part of the curriculum at The Royal College of Science in Dublin and from 1913 on it was a subject in the third year experimental science syllabus, and a final year option of natural science in the science teachers group. [14] In 1915 Cole produced a book for school teachers entitled *Ireland: the Land and the Landscape*. He suggested that naturalists' field-clubs should organise geographical study groups, taking counties or townlands as units of investigation and using Ordnance Survey maps as the basis for their projects. An active member of the British Geographical Association, Cole was elected president in 1919 and one year earlier he was instrumental in founding an Irish affiliate, the Irish Geographical Association, over which he presided from 1918 to 1922. [15] During these years he organised lectures, field excursions, and a summer school for teachers. He proposed a survey of County Dublin, and worked in vain to get geography established at Trinity College Dublin. [16]

Cole was a staunch internationalist who believed in the indissoluble unity of the British Isles, as well as the unity of physical and human geography. He claimed that geography was no longer a purely descriptive subject, but rather a challenging, analytical discipline concerned with the interrelation of spatial phenomena – a subject where geologist, historian, meteorologist and even psychologist could meet. In terms of cognitive style, he seemed to hold a view of the earth as *mosaic of patterns*. [17] Yet he did not hesitate in responding to the wider challenges of his time and place, viz. , Ireland's identity in the twentieth century. The inclusion of geography in Irish universities, he believed, could provide an antidote to what he perceived as Irish parochialism and as a measure that would assist the integration of the new Irish state into the global community of nations. [18] In this respect he echoed the organicist tone of many geographical writings of the time, even that of Kropotkin and Reclus. The concluding paragraphs of *Ireland: Land and Landscape* (1915) offer insight into these broader perspectives on Ireland:

> This little book is not going to end with a reference to the price of coal or of any other article. There are things in Ireland far more lasting than a hundred pounds in the bank or a good position in the market square. When we look at the hills and valleys, we cannot help thinking of the men and women who knew them also in the old times, who gathered wood in the forests to light their fires in ringed encampments, and who heard the wind blow round them from the sea, while they said to one another, 'Here we have made a home. '
> The tales of the old folk, long before history was written down, are full of the names of mountains and pleasant plain-lands, and of green raths placed

above the streams. There was much burning of homesteads, harrying of cattle, and clamour of heroes meeting at the ford. But the clearing of the savage woodland, the first tilling of the soil, the hands-grasp of friends, and the gracious arts of intercourse – these things, scarcely noticed by historians, were moulding the fellowship of a people. Family by family, age by age, an Irish race has grown in Ireland. Through her eastern gate and her Atlantic harbours her people have gone out across the world; but they look back to the green hillsides, to the white rivers widening into lakes, and the sweet air of the moorland, soft with rain. This book is written for those who learn and work in Ireland, and what they love best in her they will read into it for themselves. [19]

Five years later, in his best remembered work written during those turbulent war years, *Ireland the Outpost* (1920), he argued that the notion of a pure Irish national stock was a myth: Ireland was an outpost of Europe and her people were an amalgam of ethnic waves from the east:

> It is entirely normal for nations to harbour resentment toward other nations, but a nation usually has many neighbours ... In the case of Ireland such diffusion of national sentiments has been inhibited by a simple fact of physical geography: since the twelfth century, the Irish have had to settle their external differences with one power only, whose territory, like a huge breakwater, divides them from the continental turmoil to the east. [20]

Cole died in 1924 and the Geographical Association only survived him by four years. Yet when another group assembled in 1934 to found the Geographical Society of Ireland there were many who had been members of the former body and whose geographical interests had been shaped by Cole's enthusiasm. Geology, and the burgeoning field of geomorphology, provided a sound basis for understanding Ireland's landscapes and mindscapes. His joint book with Timothy Hallissy, *Handbook of the Geology of Ireland* (1924) remained the standard text in the field for thirty years. [21]

The second major foundation for modern geography in Ireland, as John Campbell has described, was commerce. [22] Commercial geography flourished particularly in Northern Ireland. Liberal Unionist industrialists and merchants urged educational reforms after the current English, Continental and American models. In Dublin, however, only geology claimed status as subject at third level. But there was much ado about educational curricula for primary and secondary levels. Ideologically-based controversy, exaggerated by political events, would mark the first quarter of the twentieth century.

In this context, two particular voices, neither of them from the ranks of 'professional' geography, are particularly fascinating. Sir Horace Plunkett (1854–1932), founder of the Irish co-operative movement, and Father Thomas Corcoran, S. J. (1872–1943), Professor of Education at University College Dublin, both addressed the issue of how relevant was the existing corpus of educational material for the understanding of Ireland's landscapes and life, the former implicitly, the latter explicitly. Both of them objected to the content and biases of texts imported from England; both had travelled widely and sought to re-shape the training of rural youth toward authentic national development. Despite these common denominators, their scenarios for training in geography and rural development were quite divergent.

SIR HORACE PLUNKETT (1854–1932)

Horace Plunkett, best known in Ireland as founder of the co-operative movement, had a keen interest in education at all levels. Schooled in the radical liberalism of the 1870s in Oxford, his travels in Scandinavia and North America stirred his enthusiasm for education toward self-help and economic development. He was a friend of Theodore Roosevelt and imbibed much of the progressivist spirit of late nineteenth century America. The root cause of Ireland's backwardness, he felt, was not so much British rule, but rather the deficiencies in Irish education. For the rehabilitation of Irish economic life and the fostering of courage, initiative and self-reliance Plunkett encouraged his Belfast friends to extend 'the economic thought with which you are indoctrinated . . . to the rest of Ireland'. [23] While the Home Rule efforts were floundering and Conservative government re-instated in England, Plunkett convened a committee of enquiry into Irish affairs during the parliamentary recess of 1895–96. One outcome of this Recess Committee's report was the setting up of a Department of Agricultural and Technical Instruction (DATI) which was to promote improvements in agriculture through the development of co-operatives, and the training of rural youth.

The modernisation of agrarian life in Ireland was Plunkett's central aim. Toward this end, of course, the question of education was vital. The educational system in Ireland at the turn of the twentieth century, Plunkett argued, 'geared to the multiplication of clerks and professional men with a distinct distaste for any industrial or productive occupation' should be supplemented by vocational education better attuned to Ireland's needs:

> . . . I believe that by awakening the feelings of pride, self-respect, and love of country, based on knowledge, every department of Irish life will be invigorated . . . In the Gaelic revival there is a programme of work for the individual; his mind is engaged, thought begets energy, and this energy vitalises every part of his nature. This makes for the strengthening of character, and so far from any harm being done to the practical movement . . . the testimony of my fellow workers, as well as my own observation, is unanimous in affirming that the influence of the branches of the Gaelic League is distinctly useful whenever it is sought to move the people to industrial or commercial activity. [24]

In his search for a suitable model, Denmark's *Folkhögskolor* offered promise. There he found programmes involving courses on history, geography, and technical skills for rural adults. Plunkett was impressed with Grundtvig:

> It is to the 'High Schools' [Högskolor] founded by Bishop Grundtvig, and not

to the Agricultural schools that the extraordinary national progress is mainly due. [25]

His Danish friends explained that their success was not due to the technical training provided, but rather to the humanities. Plunkett would add 'the nationalities': 'for nothing is more evident to the student of Danish education ... than that one of the secrets of their success is to be found in their national basis and their foundation upon the history and literature of the country'. Would Irish rural people not benefit from these lessons?

Geography as 'home area studies' had begun to attract quite an enthusiastic following in several European countries during the early years of the twentieth century – in stark contrast to the situation in Ireland, despite Cole's heroic efforts. From Balkans to Baltic, Catalonia to Norway, *Heimatkunde* (home area studies) had become one of the discipline's greatest appeals: students from many branches of science and humanities queued up for summer courses in survey, analysis and mapping of their territories. [26] Its social (and political) relevance, had taken diverse forms throughout northern and Saxon lands in the first quarter of the twentieth century, from the eliciting of national consciousness in peripheral and newly-emerging states, to the legitimation of expansionary geopolitics at the core of ambitious empires. [27] Just as the ethos of Boy Scouts training in the conquest of nature served to build character for imperial service in Britain, so indeed would the building of cooperatives in Ireland serve to conquer economies of scale and market niche in a commercially-competitive world.

Plunkett's vision, while drawing inspiration from the organicist view, was strongly oriented toward eighteenth-century Enlightenment mechanistic views of economic and social development. The products of his curriculum would constitute 'an industrial army ... organised and disciplined for the task of doing battle for Ireland's position in the world market'. [28] And this would require a constitution 'so skilfully contrived that it will harmonise all the interests involved'. [29] Unlike the organicist interpretations of areas and people, where issues of community and cultural identity played a central role, the mechanist interpretation laid emphasis primarily on economic productivity and expansion:

> The principle of agricultural cooperation with its economic advantages will, as time goes on, be further extended by the combined action of societies ... Federations, too, are being formed, as societies find that their business can be conducted more economically, for example, in dairying by centralising the manufacture of butter, or in the egg export trade by the alliance of many districts to enable large contracts to be undertaken. [30]

On the institutional front, the Department of Agricultural and Technical Instruction (DATI) took steps to promote programmes in commercial geography. In 1899 the Royal College of Science for Ireland was adumbrated under its aegis and steps were taken to improve the quality of science teaching in secondary schools. [31] Arguing along the lines previously charted by the Royal Dublin Society, it supported the setting up of a Faculty of Commerce at Queen's University Belfast in 1910. A 'Higher Certificate in Geography' for secondary school teachers was announced in DATI's *Journal* 1917–18, and set of textbooks initiated by his Assistant Secretary for Technical Instruction, George Fletcher. The editorial in this five volume *Provinces of Ireland* 1921–22 presents the case rather defensively:

> The aim of this series is to offer, in a readable form, an account of the physical features of Ireland, and of the economic and social activities of its people. It deals therefore with matters of fact rather than with matters of opinion; and, for this reason, it has happily been found possible to avoid political controversy . . . The progress and status of Ireland as an agricultural country are recognised and acknowledged, but her industrial potentialities have, until recently, been inadequately studied . . . It is hoped that these pages may contribute to the growing movement in the direction of industrial reconstruction. [32]

In Dublin geography was taught in the Rathmines School of Commerce, an institution strongly supported by the Urban District Council: the first three principals of the school had commercial geography as their teaching speciality. C. H. Oldham (1904–1909) moved to University College Dublin and became Professor of Commerce there until 1925. His successor A. Williamson (1909–1919) authored a much used text; A. G. T. Clampett (1919–1951) secured advanced certificates in commerce and commercial geography from the London School of Economics in the late 1920s. John Campbell claims that the developments at Rathmines and University College Dublin may, in fact, have spurred the decision of Trinity College Dublin to start its own school of commerce in 1925. [33]

In 1926 the Royal College of Science was absorbed into University College Dublin. Its graduates, many of whom were English, had nearly all returned home. Though the battle ground for geography was not yet at university level, it was from the National University that the most vocal opposition to all unionist-inspired texts emerged: the training of teachers and the editing of textbooks in this newly-independent state was to be moulded along strictly nationalistic and denominational lines.

TIMOTHY JOSEPH CORCORAN, S. J. (1872–1943)

Father Timothy J. Corcoran, Professor of Education at University College Dublin 1909–1942, and close associate of Eoin MacNeill, Minister of Education 1922–25, was one of the most influential architects of educational policy in the Irish Free State. The principal duty of an Irish educational policy, MacNeill had argued in the *Irish Statesman*, should be 'the building up of an Irish civilisation', for 'Irish political freedom without Irish nationality is not worth one drop of ink'.[34] Geography could be a vital shaper of such a vision. Inspired by the organicist views of historians such as von Herder and Michelet, MacNeill stressed the importance of the native language, literature, and folk traditions. Without its native language, Ireland would become 'a mere geographical expression'.[35] Geography, Corcoran claimed in 1915, was the best possible field in which the Irish language could be fostered. 'No other subject of the whole course of general education is as suitable as Geography for training in a new vernacular.'[36] As in France, Corcoran believed that geography's closest ally should be history.[37]

Corcoran, however, was quite agnostic about the liberal view in Irish universities – that education should involve more than technical training for employment. He seems to have been quite a practical man, eager to cultivate geography as a vital ingredient in 'education for the land' in rural schools particularly.[38] His views on rural development in Ireland were quite different from those of Plunkett. He harangued against those curricular reforms promulgated by the DATI which 'stretched out its grasp towards secondary and primary schools' seeking 'absorption and control, not adjustment of studies'.[39] Not only had the teaching of geography through English, by teachers of English literature, brought the subject into disrepute in secondary schools, but the texts were misleading for Irish youth:

> The Geography manuals too often used in Ireland have descended from the text-books devised fifty years ago in English, when the doctrines and the point of view of the Manchester School of Economics widely prevailed. If the eyes of its writers, thinkers, teachers, it was the big factory town that mattered. Its growing population was a clear index of progress and happiness. The countryside, where food was produced, counted for little: in England, this was accentuated, because the factory class in the State was set against the agricultural population, and wanted imports of food rather than home farming. Ship-building centres were 'featured' in school geographies: they were analogous to factory cities. The egg-producing population passed unnoticed, though, as in the comparison of Belfast with the rest of Ireland recently made, egg-production in Ireland does more for national income, for suitable human activity, and for widespread service, than the 'massed' industry of ship-building. The rural industries of all modern countries are politically less influential than those of urban areas . . . There is less place for the rural element in the geography text-book, which unconsciously reflects political and organised humanity. In it the things that have mattered are towns, cities, massed activities. For the good of the Irish people this policy should be reversed. A regional treatment of the country, directed towards the land and its uses, can be made, on the basis of structural geography, to yield abundance of varies and practical matter.[40]

Corcoran, echoing indeed some of the concurrent anti-urban bias within French geography of the 1920s, was adamantly opposed to the biases of the 'Manchester School':

> That growth of population in a city can be a national loss, may even be a distinct symptom of national disease, can easily be shown from the case of Dublin . . . For all these issues . . . the Manchester School, real authors of the teaching of Geography in English, had one judgement process. Concentrated

population counted: mass production impressed; big financial returns, represented by the movement of trade in a port, signified human well-being and civilisation. The ultimate insolence came in discourses on the infant mortality of Ireland, though the infant death rate of Dundee, for instance, vastly exceeded that of Donegal, as did that of Manchester surpass that of Mayo. [41]

If a foreign standard were desired, then Denmark, Holland, or Switzerland might provide more durable guidance. However, it would be 'Much better if we develop our own scale of human and economic values and apply them in teaching Geography no less than in handling History'.

Corcoran had a keen interest in history. He was a founder member of the Irish Manuscripts Commission (1928) and took particular interest in the publication of *The Civil Survey 1654–56*; he planned to publish existing *Early Maps of Ireland*. [42] His was certainly one of the most powerful influences on Irish geography teaching programmes at the secondary school level.

The *impasse* between 'Plunkett' and 'Corcoran' views of geography obviously reflected conflicting ideological positions on Irish independence. But there were also fundamental differences in the 'root metaphors' of these two champions of geography and rural development. Plunkett wrote in the powerfully-integrated root metaphors of organicism and mechanism; Corcoran, while admiring, e. g. , the French geographers' organicist style, wrote about geography in a contextualist style – students should focus on their own (local and national) realities first, learn about their own history and culture, and liberate themselves from views which he deemed inappropriate. The other main difference in their rhetorical styles is that Plunkett emphasised 'freedom to' develop rural Ireland, Corcoran emphasised 'freedom from' all that he deemed alien ideas. His system thus was one which would 'don the mantle of the oppressor': just as geography has been previously taught and examined in the framework of English language and literature, so in future it would be subsumed under the teaching of Irish language and life.

In hindsight it appears that Plunkett may have lost the battle on the curriculum for geography, but eventually won the war on modernisation within Irish agriculture. Corcoran won the battle for geography's place in the school curriculum, championing its integrity by insisting on its value in bolstering the spirit of Irish nationhood and self-confidence among rural schoolchildren. The emphasis on historical and cultural aspects, however, may have left insufficient place for study of contemporary economic and social questions. Winning the war on curriculum, however, was in large part effected through the training of teachers. The record of selection procedures for candidates seeking places in teacher-training institutions during the post-independence period also reveals a bias toward the Gaeltacht. [43] The school texts which best fitted the interests of the Education Ministry of this newly-independent state was the one written by the disciple of Cole, Eleanor Butler's *Structural Geography of Ireland*

(1924)[44] and the *Irish Student's Geography* (n. d.) which was prefaced as follows:

> IRELAND – OUR NATIVE LAND
> Ireland is a land shrined in song and story. Poets have sung her, calling her 'a rich and rare land', 'the land of the pure and the free', 'a land of eternal youth'. They have personified her as 'Dark Rosaleen', as 'The Shan Van Vocht', as 'Cathleen Ni Houlihan'. Of yore they sang of her as Banba, as Fodhla, as Eire – three fair queens of the Celtic imagination; and today, as centuries ago, they still revere her, addressing her as Inisfail, the Isle of Destiny.
>
> This land, which has called forth so much song, and which has a literature, the most ancient and the most wonderful in Europe, has moreover, since the dawn of history, nursed generation after generation of brave men and good women; so much so that the name by which our fathers most loved to call their land was 'Eire Ogh', that is, 'Holy Ireland'.
>
> This is the land which we propose to study in these pages, and the study will be for us a task of love; for Ireland is our own land – the dearest land on earth to us; and so any study that helps us to know more about that land cannot be without interest for us.

Butler, a friend of de Valera, and whose books were translated into Irish, might be regarded as ultimately the most influential 'gatekeeper' on school geography in post-independence Ireland. Her works incorporated at once both the spirit of the nation and provided a compendium of relevant information, all in one volume: something which had not been attempted since the eighteenth century. While other texts, authored in Britain and paying scant attention to Ireland, were widely used, Butler's texts were regarded as the most appropriate for the teaching of geography in Irish schools. In 1934, the World Federation of Education Associations met in Dublin and the teaching of geography was discussed. That same year the Geographical Society of Ireland was founded, and Eleanor Butler was a founder member.

It was in fact from teachers, who needed textbooks and pedagogical materials that the impetus came for establishing University courses in the subject. While courses were offered at the Queen's University of Belfast by the late 'twenties, and a diploma course at Trinity College Dublin since 1930, it was not until 1936 that a post in 'Geography and Education' was announced there and T. W. Freeman from Manchester was appointed. At University College Cork, a joint lectureship in geography-geology was set up in 1932 and this enabled students to follow courses leading to B. A. degree. C. S. O'Connell, appointed in 1940, assumed responsibility for geography and geology in 1946, and introduced geography as a degree subject in 1948–49. In 1940 the teachers made an unsuccessful plea to University College Dublin to begin a degree course in geography, but it was not until 1950, through renewed pressure from the Geographical Society of Ireland, that a lectureship in geography was established there in 1950. T. Jones Hughes, from Aberystwyth, was appointed. In 1959 the first chair of geography in the National University of Ireland was created at University College Cork, with C. S. O'Connell as professor. One year later University College Dublin followed suit. In 1962 a statutory lectureship in geography was set up within the geology department at University College Galway. In the same year degree courses in geography were instituted at Magee University College in Derry. In 1967 a professorship in geography was established at the New University of Ulster in Coleraine.[45] By the late 1960s geography claimed its territory at tertiary level in twelve institutions in this island. [46]

GEOGRAPHY IN IRELAND 'TWIXT TWILIGHT AND DAWN

The status of geography as school subject or as academic field has mirrored the fluctuating fortunes of nations and empires, flowering at moments of societal self-confidence, altruism, or expansionary challenge; relaxing in routine-operational tasks of imperial housekeeping and inventory, text-book writing, and functionally-specialised research in times of stability or depression. Moments of national independence offer especially good foci for examining relationships between geographical thought and context. Declarations of national independence are quintessentially geographical statements: they proclaim changes not only in the world map but also in the geographical self-understandings of people and places. How such events resonate with changing currents of academic geography varies also through time and place. Geographers have generally sought relevance to particular social contexts but they have also sought active participation in the (international) scholarly community. And chronologically speaking, the changing trends in political and intellectual life have only rarely been synchronous.

The story of geography in Ireland during that 'twilight' period spanning the first quarter of the twentieth century illustrates many of these general patterns. As elsewhere among newly-emerging states, the shaping of school curricula became a task quite as urgent as the establishment of university-level courses in the field. As elsewhere, too, the effort to liberate oneself from dependency on imported expertise and disciplinary orthodoxies deemed 'foreign' led to an emphasis on history, language and culture of the home nation. Unlike other post-colonial settings, however, there was little space accorded to geography as university field. Only through continued reliance on imported expertise, in fact, did the subject achieve a place as academic subject.

After the stormy debates of the pre-independence period and the successful establishment of school geography, it is ironic that it took at least two decades before the subject was accorded a place in university curricula. The 'founding fathers' of twentieth century geography at university level in Ireland – Estyn Evans at Queen's, T. W. Freeman at Trinity, C. S. O'Connell at University College Cork, and T. Jones Hughes at University

College Dublin – faced a herculean task. Initially unassisted, they tried to offer a broad range of subjects spanning both bio-physical and human aspects of the discipline and to lay the foundations for research on Irish topics. While differing in intellectual style, all four fostered links with history; for three at least a favoured research focus was on landscapes and life in pre-twentieth century contexts.[47] Evans especially cultivated the trilogy of humanities-geography, anthropology, and history.[48] Freeman supported traditional regional geography which included both physical and human branches of the field.[49] At University College Cork, Professor Charles O'Connell lectured on all branches of the field to students from several faculties, and always sought to maintain an integrated and environmentally-sensitive field. Jones Hughes cultivated historical and social geography, strengthened the Welsh connection which continues to this day.[50]

That three of the four 'founders' were themselves Nonconformists, and that the majority of subsequent posts were occupied by either their own students or by others from Britain may also be significant. The cognitive styles of geography taught and fostered at university level during the 1930s, 40s, and 50s in Ireland mirrored the predominantly chorological emphases in the discipline internationally: the dispersed views of *world as mosaic*, or *world as arena of spontaneous events*.[51] Prudently perhaps, emphasis was placed on objective scientific observation and analysis, on careful scrutiny of historical sources, on field training and mapping rather than on claims for relevance for practical applications of the subject within the concurrent societal context. The strength and quality of undergraduate training in Ireland, north and south, has been applauded by colleagues in Europe and North America, where many of the graduates have pursued their careers.

Looking at the geographical thought in popular and political contexts, there appears to be a strong contrast between the rhetorical styles of pre- and post-independence Ireland. The prospectus for Irish nationhood, as proclaimed by Thomas Osborne Davis in the mid-nineteenth century, revealed an ecumenical and Romanticist vision:

> A Nationality which will not only raise our people from their poverty, – by securing to them the blessings of a domestic legislature, but influence and purify them with a lofty and heroic love of country – a Nationality of the spirit as well as the letter – a Nationality which may come to be stamped upon our manners, our literature, and our deeds – a Nationality which may embrace Protestant, Catholic, and Dissenter, Milesian and Cromwellian, – the Irishman of a hundred generations and the stranger who is within our gates – not a Nationality which would prelude civil war, but which would establish internal union and external independence – a Nationality which would be recognised by the world, and sanctified by wisdom, virtue and time.[52]

Padraic Pearse, leader of the 1916 Rising, involved elements of both Enlightenment and Romanticism:

> A free Ireland would not, and could not, have hunger in her fertile vales and

squalor in her cities. Ireland has resources to feed five times her population; a free Ireland would make those resources available. A free Ireland would drain the bogs, would harness the rivers, would plant the wastes, would nationalize the railways and waterways, would improve agriculture, would protect fisheries, would foster industries, would promote commerce, would diminish extravagant expenditure (as on needless judges and policemen), would beautify the cities, would educate the workers (and also the non-workers, who stand in direr need of it), would, in short, govern herself as no external power – nay, not even a government of angels and archangels could govern her.[53]

Post-independence Ireland was to become anything but that globally-minded Nation of which Thomas Davis, George Russell, and even Pearse had dreamed. In the words of the historian Terence Brown, it became rather:

> . . . a petit-bourgeois state expressing the prudent and inhibiting values of farm and shop which allowed the children of the nation to bear the brunt of its ostensibly most adventurous policy, the revival of a language, whilst the opportunities implicit in that independence so eagerly sought went begging in a fairly general acquiescence in comfortably provincial modes of social life and art. The economic realities of independent Ireland were against any great national resurgence . . . Too many thought a respectable survival was enough.[54]

Pearse had sounded the clarion before the 1916 Rising that challenged a nation to economic, social and cultural deeds of daring. In 1928 George Bernard Shaw issued an equally demanding warning:

> Ireland is now in a position of special and extreme peril. Until the other day we enjoyed a factitious prestige as a thorn in the side of England, or shall I say, from the military point of view, the Achilles heel of England? . . . when we were given a free hand to make good we found ourselves out with a shock that has taken all the moral pluck out of us as completely as shell shock. We can recover our nerve only by forcing ourselves to face new ideas, proving all things, and standing by that which is good . . .
>
> The moral is obvious. In the nineteenth century all the world was concerned about Ireland. In the twentieth nobody outside Ireland cares twopence what happens to her. If she holds her own in the front of European culture, so much the better for her and Europe. But if, having broken England's grip of her she slips back into the Atlantic as a little grass patch in which a few million moral cowards cannot call their souls their own . . . then the world will let 'these Irish' go their own way into insignificance without the smallest concern. [55]

The twentieth century post-independence scene in Ireland does seem to fit the central thesis of this essay: pre-independence models of a global, integrative, nature, post-independence models of a housekeeping, domestically-oriented nature. Clifford Geerz identified a tension between two impulses in newly independent nationalist states in this century.[56] The one [*essentialism*] is an impulse to answer the question 'Who are we?' by employing 'symbolic forms drawn from local traditions'. Through these the new state can 'give value and significance to the activities of the state, and by extension to the civil life of its citizens'. The other [*epochalism*] impulse is to answer the question of national identity through discovering the 'outlines of the history of our time and what one takes to be the overall direction and significance of that history. ' This tension, it has been noted, was really evident in the Irish situation.[57]

Concluding Queries

To conclude, then, with an invitation for dialogue, particularly with colleagues who face a post-independence situation for the discipline in their respective countries, on the following claims:

Integrated world views have provided rhetoric for movements leading to national autonomy: those of *mechanism* salient in the framing of States, those of *organicism* more salient in bolstering the Nation as symbol of cultural identity, territory and relationships to Nature. In post-independence times there has been a stabilisation of disciplinary practice around issues of domestic order, resource inventory, and boundary definition. The root metaphor of world as *mosaic of patterns* has served geographies of *status quo*, while that of world as *arena of spontaneous events* has provided eloquent critique of the taken-for-granted.

Paradigm shifts in geography may be associated with analogous trends. Once freedom has been won, and the new paradigm is firmly in place, the discipline has often witnessed a stabilisation of endeavour, an intellectual plateau, where academic interests settle around issues of international housekeeping, dismissing global horizons. The creative scholars, after some critical reflection, would move on to new intellectual frontiers.

Contextually speaking, while the 'national interest' may provide an integrative framework for geography's educational role in post-independence times, the often divergent interests of economy and culture may lead to a dispersed array of functionally specialised pursuits, e. g. , systematic enquiry into resources, population, economic development and trade on the one hand, or historically-based regional studies on landscape, culture and identity on the other.

The rhythms of political and scientific 'revolutions' have rarely been synchronous. The periods during which gatekeeping has played its most decisive role have been those twilight ones when intellectual energies address issues of liberation from previous tyrannies. It is at such periods that a wise balance be maintained between the rhetorics of 'freedom from' and those of 'freedom to'; what matters most, both intellectually and politically, today as well as yesterday, is that the 'dawn' may hear a plurality of voices each acknowledging the on-going challenge of fashioning appropriate geographies for the ever changing realities they seek to elucidate.

NOTES

1. The substance of this paper has been published in a longer article entitled 'Gatekeeping Geography through National Independence: Stories from Harvard and Dublin', *Erdkunde* 49(1): pp. 1–16.
2. P. Kropotkin, 'What Geography ought to be', *The Nineteenth Century* (December 1885): p. 942.
3. These questions are addresssed in a forthcoming volume edited by A. Buttimer and S.D. Brunn, *Text and Image: Constructing Regional Knowledges*, which contains case studies from nineteen countries.
4. The expression is used in a very general sense to cover the record of geography as scholarly field of enquiry into relationships between humanity and environment, of geography as academic discipline and school subject, and also that of taken-for-granted geographical thought expressed in policy and planning, popular press, and everyday public life.
5. G.H. Davies, 'The Making of Irish Geography. II: G.A.J. Cole 1859–1924', *Irish Geography*, 10 (1977): pp. 90–94.
6. J. Campbell, 'Modernisation and the Beginnings of Geography in Ireland', Presentation at International Geographical Union Commission on the History of Geographical Thought, Utrecht, August 1991, forthcoming in H. van Ginkel and V. Berdoulay, eds., *Geography and Professional Practice*, special issue of *Nederlandsche Geographical Studies*.
7. These 'root metaphors' are elaborated in A. Buttimer, *Geography and the Human Spirit* (Baltimore, 1993).
8. G. Fahy, 'Geography in the Early Irish Monastic Schools: a Brief Review of Aaribheartach MacCosse's Geographical poem', *Geographical Viewpoint*, 3 (1974) 31–43; and 'Geography and Geographic Education in Early Nineteenth Century Ireland', *Proceedings of the Educational Studies Association of Ireland* (1980): pp. 423–443.
9. H.F. Berry, *A History of the Royal Dublin Society* (Dublin, 1915); T. deVere White, *The Story of the Royal Dublin Society* (Tralee, 1955). See also Campbell, 'Modernisation and the Beginnings of Geography in Ireland'.
10. G.L.H. Davies and R.C. Mollan, eds., *Richard Griffith 1784–1878* (Dublin, 1980).
11. G.L.H. Davies, *North from the Hook. 150 Years of the Geological Survey of Ireland* (Dublin, 1995).
12. P.N. Wyse Jackson, 'On Rocks and Bicycles: a Biobibliography of Grenville Arthur James Cole (1859–1924) Fifth Director of the Geological Survey of Ireland', *Geological Survey of Ireland Bulletin*, 4:2 (1989): pp. 151–163; Davies, 'The Making of Irish Geography. II'.
13. W.M. Davis, 'A Geographical Pilgrimage from Ireland to Italy', *Annals of the Association of American Geographers*, 2 (1912): pp. 73–100.
14. Campbell, 'Modernisation and the Beginnings of Geography in Ireland'.
15. G.A.J. Cole, 'The Irish Geographical Association: Presidential Address', *Geographical Teacher* 10 (1919–20): pp. 93, 276–9.
16. See Campbell, 'Modernisation and the Beginnings of Geography in Ireland'.
17. Buttimer, *Geography and the Human Spirit*, ch. 4.
18. G.L.H. Davies, 'This Protean subject. The Geography Department in Trinity College Dublin 1936–1986', Dublin: Trinity College Dublin Department of Geography, 1986.
19. G.A.J. Cole, *Ireland: The Land and the Landscape* (Dublin: The Educational Company of Ireland, 1915), p. 140.
20. Davies, 'The Making of Irish Geography. II', p. 93.
21. *Ibid.*, p. 91.
22. Campbell, 'Modernisation and the Beginnings of Geography in Ireland'.

23. H.C. Plunkett, 'Address at the Queen's College', *The Irish Homestead*, 4 February 1899: pp. 89–90.
24. H.C. Plunkett, *Ireland in the New Century* (London, 1904), p. 129.
25. *Ibid.*, p. 131.
26. A. Buttimer, 'Edgar Kant and *Heimatkunde*: Balto-Skandia and Regional Identity', in D. Hooson, ed., *Geography and National Identity* (Oxford, 1994), pp. 161–183.
27. In terms of cognitive style, *Heimatkunde* incorporated at least three distinct world views, or root metaphors. The mapping of phenomena and inventory of a landscape's history and resources yielded a *mosaic* of patterns which could eventually constitute sheets for the national atlas, to be later perused in the interpretation of places, cultural traditions, and regional life. There were two others, however, which vied for priority in the writings of Plunkett: world as *organic whole* and world as *mechanical system*.
28. H.C. Plunkett, *DATI Journal* (1901–2): p. 692.
29. Plunkett, *Ireland in the New Century*, p. 183.
30. *Ibid.*, p. 133.
31. Campbell, 'Modernisation and the Beginnings of Geography in Ireland'.
32. G. Fletcher, ed., *The Provinces of Ireland* (Cambridge, 1921), p. v.
33. Campbell, 'Modernisation and the Beginnings of Geography in Ireland'; see, however, Davies, 'This Protean Subject'.
34. E. MacNeill, 'Irish Educational Policy', *The Irish Statesman*, 5 (1925): pp. 169.
35. D. McCartney, 'MacNeill and Irish Ireland', in F.X. Martin and F.J. Byrne, eds., *The Scholar Revolutionary: Eoin MacNeill 1867–1945 and the Making of the New Ireland* (Dublin, 1973), p. 79.
36. T.J. Corcoran, 'Geography as a Subject of General Education in Ireland', *Studies*, 12 (1923): pp. 616–622, p. 618.
37. *Ibid.*, p. 622; see also Campbell, 'Modernisation and the Beginnings of Geography in Ireland'.
38. Corcoran, 'Education for the Land in Ireland', *Studies*, 4 (1915): pp. 351–356.
39. Corcoran, 'Geography as a Subject of General Education in Ireland', p. 617.
40. *Ibid.*, pp. 619–20.
41. *Ibid.*, p. 620.
42. D.F. Gleeson et al., 'Father T.J. Corcoran, S.J.', *Studies*, 32 (1943): pp. 153–162.
43. N. Johnson, 'Nation-Building, Language and Education: The Geography of Teacher Recruitment in Ireland, 1925–55', *Political Geography*, 11: 2 (1992): pp. 170–189.
44. E. Butler, *Structural Geography of Ireland* (Dublin, 1929).
45. R.E. Glasscock, 'Geography in the Irish Universities', *Irish Geography* 5:5 (1967): pp. 459–468.
46. Glasscock, 'Geography in the Irish Universities'; J.E. Killen and W.J. Smyth, *Bibliography of Irish Geography 1987–1991* (Dublin, 1992), pp. 1–6.
47. G.L.H. Davies, 'Thomas Walter Freeman and the Geography of Ireland – a Tribute', *Irish Geography*, 6 (1973): pp. 521–528 and *Irish Geography: The Geographical Society of Ireland Golden Jubilee 1934–1984* (Dublin, 1984); J.A. Campbell, *Geography at Queen's. An Historical Survey* (Belfast, 1988); W.J. Smyth and K. Whelan eds., *Common Ground. Essays on the Historical Geography of Ireland* (Cork, 1988).
48. E.A. Evans, *Irish Heritage* (Cambridge: Cambridge University Press, 1928).
49. T.W. Freeman, *Ireland: Its Physical, Historical, Social and Economic Geography* (London: Methuen, 1969 [1950]).
50. Smyth and Whelan, eds., *Common Ground*; see, for example, J. Andrews, 'Jones Hughes's Ireland: A Literary Quest' pp. 1–21.
51. See Buttimer, *Geography and the Human Spirit*, chapters 3 and 6.

52. Cited in C. Maxwell, *A History of Trinity College Dublin 1591–1892* (Dublin, 1946), p. 220.
53. P.H. Pearse, *Political Writings and Speeches* (Dublin, 1952 [1916]) p. 180.
54. T. Brown, *Ireland: A Social and Cultural History 1922–79* (Glasgow, 1981), p. 136.
55. G.B. Shaw, Note in *Irish Statesman*, 17 November 1928, p. 208.
56. C. Geerz, *The Interpretation of Cultures* (London, 1975), pp. 234–54.
57. Brown, *Ireland: A Social and Cultural History*; J.L. Lee, *Ireland 1912–1985* (London, 1991).

Chapter 10

Science and the Cultural Revival: 1863–1916

Sean Lysaght

That science should figure as an element in the Irish Cultural Revival of the late nineteenth- and early twentieth-century appears at first view paradoxical, since many of the premises of that revival were anti-scientific. The familiar anatomy of the movement led by Yeats, Gregory and Synge, with its roots in romantic and symbolist anti-materialism, need not be reiterated here. Yet this simplification, like any simplification, invites scrutiny.

The main thrust of cultural criticism from within academic circles in recent years has been to point out how essentially fictive or ideological the multitude of images of Ireland were which proliferated in the decades in question; how these images were not the whole or the only story of lived experience in the community which they claimed to represent. The theoretical or ideological source of the majority of these commentaries is historicist; at the same time, concealed beneath the historicism, there is a strong current of cultural nationalism which resents the virtual monopoly enjoyed by Protestant Ireland of Irish image-making during the Cultural Revival. With sympathies such as these, this tradition of criticism is quick to serve those who were dispossessed or misrepresented by politics, literature or economic organisation. If we consider that science was, in most of its manifestations, another aspect of power, authority and respectability, it hardly comes as a surprise to discover that it finds few sympathisers among those writing from within the professional field of cultural criticism.

Prior to a number of pioneering articles published in the last ten years by writers such as John Wilson Foster and Dorinda Outram,[1] there was a curious silence about Irish science *as a component of culture*. (Apart from the local Irish factors which operated in the case in question, this silence was no doubt also a natural outcome of a wider division of disciplines which has only recently begun to be eroded by the rise of interdisciplinary studies.) It was both ironic and appropriate that the recently most widely-cited statement of Irish scientific achievement should appear in the nationalist, essentialist setting of Richard Kearney's multi-disciplinary volume: *The Irish Mind: Exploring Intellectual Traditions* (1985): this was Gordon Herries Davies' contribution on 'Irish Thought in Science'. In this article, Herries Davies writes that between 1780 and 1880, there was a thriving scientific culture in Ireland, represented mainly in Trinity College, the Royal Dublin Society and the Royal Irish Academy. The principal disciplines which are covered in his invaluable survey are geology, physics, mathematics and astronomy; natural history comes into prominence only after the 1880s, by which time the other disciplines have entered into a period of decline. Surveying the closing decades of the nineteenth century, Herries Davies marshals an impressive inventory of loss. He goes on to point out, however, that 'Only one sector of Irish science showed any real buoyancy during the closing decades of the nineteenth century: field natural history. Indeed, within that area there was a local boom – a boom in which a major inspirational figure was Robert Lloyd Praeger.'[2]

Natural history therefore presents itself as an anomaly among the other scientific disciplines. The mathematical, physical and geological sciences comprising Ireland's major century of scientific achievement (c. 1780 – c. 1880) had been promoted by the ruling Protestant elite during years of relative political robustness. At the same time, according to Herries Davies, 'new scientific institutions were developing overseas, especially in Germany and the United States, and as international science became increasingly competitive, it followed that the high prestige of Irish science had to face a growing challenge as a part of the natural course of events'.[3] A key issue here was the funding of scientific research, which was becoming increasingly expensive, requiring large outlays for plant and expertise, and in this area Ireland was poorly equipped to compete, a disadvantage it shared with England. Large, corporate laboratories with specific technical responsibilities were changing the face of science on the Continent and in the United States: 'The day of the isolated researcher was clearly limited.'[4]

While corporate scientific organisation was displacing the individual enthusiast in the second half of the nineteenth century in the 'hard' disciplines such as physics and chemistry, the same was not true in natural history. Natural history continued to attract the enlightened amateur working outside the context of an academic or scientific institution. Given that natural history collecting involved connections between localities and specimens, there was, in the years prior to the growth of ecology as a science, an environmental and local component in natural history. On the other hand, the content of the chemical, physical and mathematical sciences was international. With its particular status, free of technical implementation, natural history clearly invites cultural analysis. Indeed, the sociologist Steven Yearley considers that *all* Irish science in the nineteenth century, in the hard as well as in the 'soft' disciplines such as natural history, tended to be a *cultural*, as distinct from a purely *technical*, activity. Among the dominant Anglo-Irish elite, 'Scientific attainment might be esteemed but it would be evaluated alongside other cultural accomplishments, not in terms of its utility.'[5]

The rise of Irish natural history in Praeger's time was conditioned by the leisure, mobility and cultural aspirations of a professional, mainly Protestant middle and upper class. It coincides with the explosive growth of interest in things Irish during the last decades of the nineteenth century; geographically, it was stimulated by the expansion of the Irish railway network under the Congested Districts Board: 'few things,' wrote F.S.L. Lyons, 'contributed more to "opening up" the west.'[6] Natural history finds parallels in the folklore collecting and philological research which was carried on in other western locations at this time by Douglas Hyde, Lady Gregory, J.M. Synge and others. The main biological syntheses of this interest are Praeger's *A Tourist's Flora of the West of Ireland* (1907) and the collaborative Clare Island Survey of 1909–11. The enticement, and the imperative, of the west of Ireland as a site of imaginative resort is captured by Joyce in his story 'The Dead', where Gabriel Conroy's supreme moment of sentimental reconciliation with his country is modulated as the realisation that 'The time had come for him to set out on his journey westward.' By the 1920s, according to Terence Brown, the west and the western island had been confirmed 'as the main locus of Irish cultural aspiration'.[7]

The west of Ireland as a site of cultural exploration and imaginative investment is now well-known from the familiar branches of cultural study; natural history too deserves inclusion in this picture. The London-born botanist Alexander Goodman More, who later worked as curator at the Natural History Museum in Dublin, was won over to Ireland by a series of visits to Castle Taylor, near Ardrahan in Co. Galway in the 1850s; his 'conversion' anticipates Synge's famous voyage westward to the Aran Islands at Yeats's urging some four decades later. In collaboration with David Moore of Glasnevin, More produced the first topographically-based survey of Irish botany, *Cybele Hibernica* (1866), which was the basis for the later work of Samuel Stewart, Praeger, Nathaniel Colgan and others.[8] If botany was the best-represented branch of natural history and the one which most concerned Praeger, the leading protagonist of the movement, there was other activity on a wide front of local collecting and species determination: ornithology, conchology, entomology and marine biology all go through a phase of rapid expansion during these years. Natural history comes to prominence in Ireland at a time of excitement in many cultural areas, and is a special case for inclusion in that 'expanded sense of culture' which John Wilson Foster recently called for, identifying natural history as a phenomenon which merges 'in one direction with biology and in the other with literature'.[9]

As Herries Davies mentions, it was Praeger's career which provided the natural history movement with its driving force. Praeger was born in Belfast in 1865; his maternal family, the Pattersons of Holywood, were eminent members of that Presbyterian middle class which had prospered with the boom in Belfast's linen, engineering and ship-building industries during the nineteenth century. Prosperity brought with it both an attachment to educational ideals and the leisure to indulge in hobbies and cultural pursuits which were more reminiscent of aristocratic privileges.

Praeger's maternal grandfather, Robert Patterson (1802–72) was a founder of the Belfast Natural History and Philosophy Society (BNHPS) and the author of school text-books on zoology and of a book on the insects in Shakespeare's plays.[10] Speaking to a meeting of the BNHPS in 1840, Robert Patterson recommended the study of natural history 'Because it exercises a sanitary influence, on both the perceptive and the reflective faculties . . . and because it exerts a powerful influence on the moral and devotional character'. His son Robert Lloyd Patterson (1836–1906) may have been following his father's recommendation when he retired after a brief career in the mill-furnishing business to devote himself to hunting, navigation, natural history and art collecting. Leisure and wide attainments such as these were part of the atmosphere of the closely-knit family which Praeger grew up among, in Croft House, Holywood.

The family spent regular holidays on the coast of Antrim and in the Glens, where 'Robin's' love of nature and landscape were stimulated. In 1879, going farther afield, they travelled to Cumbria and met many English tourists in the grip of a contemporary Victorian fashion for fern-collecting. Praeger's first systematic recording of plants dates from this time. Back in the north-east of Ireland, Praeger soon became a member of the Belfast Naturalists' Field Club (BNFC), an offshoot of the older BNHPS formed in 1863. Here he met many other naturalists, including Samuel Alexander Stewart, whose *Flora of the North-East of Ireland* was published by the BNFC in 1888 [and republished by the Institute of Irish Studies in 1992]. During the 1880s it is possible to trace, through the *Proceedings of the Belfast Naturalists' Field Club*, the rapid rise of this young recruit through the ranks from being a competitor to presiding as a judge in the club's plant collection competitions.

For Praeger and his contemporaries, the natural history of their region was a *terra incognita* seeming to offer limitless possibilities for exploration. The identification and naming of landscape forms and biological specimens was buoyed up on a sense of the novelty of the enterprise: no-one had done this before. New criteria of thoroughness and hitherto neglected animal and plant groups (e.g. lichens, charophyta, foraminifera, coleoptera) held out the prospect of inexhaustible funds of data and nomenclature awaiting the diligent searcher. There is, in the scientific reports of the time, a muted euphoria which reminds one of Yeats's early journalism promoting the new cultural movement.

This scientific work was not, of course, done arbitrarily or in a vacuum: the display of flags at the field club exhibition meetings or *conversaziones* made quite explicit the political and regional context of these pursuits. The BNFC and its exhibition meetings were a cultural expression of a group within the Dissenter community which needed to define itself in several ways. Firstly, history and locality could be appropriated through the nomenclature of botany, geology, palaeography and antiquarianism; this supplied an obvious atavistic dimension for a settler community conscious of its provenance in Scotland. Secondly, and crucially, the cultural and scientific club provided a secular arena for social intercourse at a time of increasing religious polarisation. 1859, just four years prior to the establishment of the BNFC, was the year of the great evangelical revival in radical Ulster Protestantism, a development which must have alienated many enlightened spirits among the Dissenters.[11] Against a background of militant, even hysterical spiritualism, the field club functioned as a kind of freemasonry where religious discourse had no place. Sectarian antagonisms come closest to Praeger's own career in the mid-eighties when he was working as a recently-qualified engineer at the new Alexandra Dock in Belfast port: the docklands were the scene of vicious rioting in 1886 at the time of the First Home Rule Bill.

A professional career as an engineer did not offer Praeger the scope he needed as a naturalist, so eventually, after some years' freelancing and faced with the prospect of a full-time commitment to engineering, he managed to get a job as a librarian in the days when a liberal education was sufficient qualification for such a post. Praeger took up his position at the National Library in Dublin at a time of major expansion of the educational facilities in the area between Kildare Street and Merrion Square. In 1877, under the Dublin Science and Art Museum Act, the Westminster government's Science and Art Department had taken control of the Royal Dublin Society's library, its art school, its biological and scientific collections, and its Botanic Gardens. (A simultaneous attempt at amalgamating the RDS and the Royal Irish Academy was successfully resisted by the latter.) New buildings were needed to accommodate the Science and Art Department's expanding role and the National Library and Science and Art Museum were provided with new premises in 1890 which are those they occupy to this day. The Science and Art Department's control over these educational institutions was to be relatively short-lived, however: in 1899 the Agriculture and Technical Instruction (Ireland) Act returned the control of technical and scientific education to Irish hands under the newly-created Department of Agriculture and Technical Instruction, with Horace Plunkett as vice-president.[12]

The expansion of the Science and Art (now the National) Museum into its new Kildare Street premises heralded an era of recruitment and change, with many people being recruited in a consultative or part-time capacity from the Royal College of Science and Trinity College. Many of them played prominent parts in the expanding field club movement and in the Royal Irish Academy's scientific surveys, and were regular contributors to the *Irish Naturalist*. These included the Keeper of the Natural History Department, Robert Francis Scharff, and his assistant George Herbert Carpenter, who founded and co-edited the *Irish Naturalist* with Praeger in 1892. Other notable members who were wholly or partly based at the Natural History Museum during Praeger's career as a librarian in the adjacent building were Albert Russell Nichols, James Nathaniel Halbert, Matilda Knowles and Rowland Southern; their respective specialisms would be vital components of the surveys later carried out on Lambay and Clare Island. Southern especially, with his enormous contribution on marine biology, was a man whose grasp of natural history equalled Praeger's, although his chosen science had less potential than Praeger's to reach a wider audience.

Before leaving Belfast, Praeger had been enthusiastic about the spread of the field clubs from their original parent in Belfast. The Dublin Naturalists' Field Club (DNFC) had been founded in 1886; Cork and Limerick followed suit in 1892 and 1893 respectively. (Regional archaeological societies whose activities overlapped substantially with those of the field clubs were also founded at this time in Kildare (1891), Cork (1891), Waterford (1893) and Galway (1900).[13] In 1892, Praeger was trying to persuade a diffident Mrs. Leebody to establish a Derry Naturalists' Field Club: 'why should not *you* do this? It would not involve much labour, and you would be conferring a boon on Irish science, with Field Clubs at Derry, Belfast, Dublin, Cork and Limerick, we only want one at Galway and one at Sligo to make us perfect.'[14] Faced with her evident reluctance, nothing became of the proposal however.

Naturally, Praeger threw his weight behind the Dublin club as soon as he moved to Dublin. The social profile of the DNFC was quite different from that of the Belfast club and reflected the different positions of natural science in the two communities. Many of the Belfast members were middle-class businessmen; the Dublin club on the other hand comprised many members of the 'professocracy', as well as some large landowners. The chief mover in the formation of the DNFC had been the anthropologist and naturalist Alfred Cort Haddon, who deliberately modelled the DNFC on its Belfast counterpart. The founding committee of 1886 included five university professors, two botanical curators, a museum curator, a librarian, a clergyman, and three landowners. The composition of the DNFC was therefore much more genteel than its Belfast counterpart.[15] Its membership had grown from an initial ninety-nine to 160 in 1894, when Praeger undertook a complete review of the activities and prospects of the Irish field clubs, in a series of articles in the *Irish Naturalist*. At that time, the DNFC held its winter meetings at the Royal Irish Academy's house in Dawson Street, including an annual exhibition meeting or *conversazione*, while during the summer it organised field excursions to places of antiquarian and scientific interest. The greater representation of academics in the DNFC meant that the displays at the Dublin *conversaziones* were somewhat more technical, and that the whole occasion was somewhat less festive than in Belfast. Nonetheless, there was variety: the 1894 *conversazione* of the DNFC at Academy House in Dawson Street included a wide range of scientific exhibits: liverworts, spiders, insects, foraminifera, sea-weeds, insectivorous plants and molluscs. Dr. McWeeney 'exhibited a very fine series of bacilli of diseases, such as diphtheria, cholera, and tuberculosis, and explained lucidly the nature of these minute organisms'. Robert Welch of Belfast was present with photographs of a BNFC outing to Donegal, and a range of ethnographic views 'of Irish country life, illustrating the occupations, conveyances, and monuments of the people in the more remote parts of the country'.[16] The Victorian taste for natural design was exemplified on this occasion by

> a large table on which, amid living ferns and grasses, a large collection of fresh Fungi were naturally arranged as if growing on a green sward of mosses, each species bearing its scientific name. There were agarics of all shapes and sizes; boleti, puffballs of various kinds, hydnums and polypori. The whole exhibited a wonderful variety of form, and almost every shade of colour – greens, purples, scarlets, browns, yellows, and whites, and showed in a striking degree the variety and beauty of this class of plants.[17]

In an *Irish Naturalist* article on the DNFC, Praeger noted the wealth of facilities in Dublin, 'trained scientific men ... societies, schools and museums ... advantages which were almost entirely wanting when the hardworking citizens of the northern capital founded their Club'.[18] The foundation of clubs in Cork and Limerick he also considered to be 'one of the most interesting events in the recent history of Irish science'.[19] (Against this background of expansion in the south, the Belfast parent still remained by far the largest club, with the greatest range of activities. Membership increased from 250 in 1890 to almost 500 in 1894. Over 600 members and their friends attended the BNFC *conversazione* in the Exhibition Hall on 14th November 1895. The Belfast *conversaziones* were the models for all of the others, and were rivalled only by the Dublin club.) All in all, in 1894 Praeger considered that 'Here in Ireland, we cannot boast that by any means so general an interest is taken in natural science as in England; but within the last decade there has been a most gratifying and encouraging increase in the number of workers.'[20] 'Four working Clubs', he continued, 'for over 31,000 square miles of country is certainly a very modest allowance, but we must be content, and hope for better things to come.'[21] Noting the difficulties faced by the new clubs, Praeger announced the formation of an Irish Field Club Union at the end of 1894, so that the stronger clubs in Dublin and Belfast might 'encourage and assist their southern brethren'.[22] The organisation was intended to improve communication between the clubs through joint meetings and excursions, and through a lecturer exchange scheme whereby speakers from the clubs would travel to other clubs to communicate their findings. Praeger foresaw 'a true and lasting Union, a bond of sympathy and friendship and scientific intercourse that will help the Clubs in their work, and stimulate them in their forward march; a Union which will be a pillar of strength to the Field Clubs, an aid to British science, and a credit to Ireland'.[23]

The Irish Field Club Union (IFCU) held its first joint excursion between 11th and 17th July 1895 in Galway and the surrounding district. The joint party was made up mainly of the Belfast and Dublin clubs, although the Limerick and Cork clubs were also represented. A welcoming dinner on the night of their arrival was held at the Railway Hotel and was attended by the President of Queen's College, Sir Valentine Blake and Lady Blake and a gathering of local dignitaries: the High Sheriff of Galway, local military officers, clergymen and engineers were also present. On the Sunday, traditionally a day free of scheduled field-club activities, a further reception was held for the visitors at the Queen's College. Apart from these social engagements, the naturalists spent several days exploring the Galway region. On the day of arrival, they managed a climb up Gentian Hill outside Galway to take in the views of the bay; the following day, a party of over 100 travelled by train to Recess and continued from there, on foot and 'driving' across what they called 'the great plain of South Connemara'. On Saturday, they sailed across Galway Bay to Ballyvaughan, where they took in the bare grey limestone hills of the Burren, a region previously known at that time to only a handful of botanists. Here they were met by the botanist and nurseryman P.B. O'Kelly, who gives his name to the orchid sub-species. The party travelled by boat to Aranmore on Monday, and the normal schedule of outings was completed the following day with a trip to Oughterard and the western shore of Lough Corrib. On each of these trips there was, naturally enough, a core group of enthusiastic collectors; along with these came others who were less intensely motivated but who, no doubt, were glad to watch the collectors poring over their finds and discussing their results. Each evening was occupied with the sorting and drying of specimens, a phase of activity which Praeger always held to be of great value to all concerned.

Following this extremely successful occasion, the IFCU held several further triennial outings, although the elation of the first outing proved difficult to repeat. In Kenmare in 1898 a spell of extremely hot, dry weather made fieldwork onerous of not impossible, and we hear Praeger complaining that very few new finds were made. In the particular field of botany, it was the case that well-worked areas were yielding a diminishing harvest of new finds; alternatively, the criterion of novelty was becoming more technical as the seasons advanced, so that the amateur without rigorous knowledge was being alienated from the movement. In 1899 these new conditions were reflected in the remarks of the President of the BNFC when he said that 'The progress of science has been so great that it [has] become necessary for anyone who wished . . . to investigate any branch to become a specialist . . . the older race of field naturalist [is] passing away.' At field club excursions in these years it becomes apparent that a gap separated the keen amateurs with specialist knowledge from those who simply wanted a carefree day out in the country, punctuated by a little botanising, one or two visits to sites of antiquarian interest and perhaps an hour or two spent sketching. The Irish Field Club Union held its triennial excursion to Sligo in 1904 where, on the Friday evening 'the scientific proceedings in the Town Hall were suspended early in favour of dancing' – a clear instance of the specialist spirit yielding to popular demand.

The developing rift between popular excursionism and natural history led some of the scientists to look away from the field club movement towards specialist study, as was to happen in the Clare Island Survey. There is one notable instance in these years, however, of an alliance between science and social interest, in the meeting of the British Association for the Advancement of Science (BAAS), held in Dublin between 2nd and 9th September 1908. By this time, the spirit of enlightened amateurism which had led to the foundation of the BAAS was itself feeling the pressure of the new climate of specialism in science, as the chairman of the delegate conference, Professor H.A. Miers of the University of London, conceded on 3rd September in the course of his address on 'The Educational Opportunities of Local Scientific Societies', where he pointed out that the vastly improved standard of scientific literature in recent years had alienated 'the intelligent amateur': 'he can no longer get an adequate insight into the modern advances of science without either going through a course of special reading in text-books of various grades – for which he has no time – or attempting to master a treatise which he can hardly be expected to understand without a preliminary training of some sort.'[24] Notwithstanding subtle shifts of this kind, the southern capital responded with lavish hospitality to the flattery of the 'Parliament of Science' holding its annual gathering in Dublin. Praeger and Grenville A.J. Cole collaborated on a special handbook to the Dublin area for the benefit of the visitors,[25] and Praeger himself attended the meeting as the Royal Irish Academy's representative. On this occasion, a quite hectic schedule of social entertainment arranged for the scientific delegates threatened to overwhelm the various sectional conferences which constituted the work of the BAAS meeting proper. The opening of the meeting was marked by a civic reception at the Mansion House; the following afternoon a garden party was hosted at Trinity College by the Provost and the Senior Fellows. On the evening of the 3rd, the Royal Dublin Society hosted a special *conversazione* at Leinster House in Kildare Street, where a special emphasis was placed on exhibits of Irish flora and fauna.[26] Later in the week there were semi-private parties at Dunsink Observatory and St Patrick's Cathedral, and two special matinee performances of Irish plays at the Abbey Theatre, including Yeats's *The Hour Glass*, Lady Gregory's *Spreading the News* and Synge's *Riders to the Sea* and

In the Shadow of the Glen.²⁷ The last few days of the 'meeting' were marked by a series of high-profile social events. Large parties were hosted by Lord and Lady Ardilaun at St Anne's, Clontarf (7th September), Lord and Lady Iveagh (8th), and the Lord Lieutenant and Lady Aberdeen at the Viceregal Lodge (9th). On Tuesday 8th there was also a daytime entertainment at the zoological gardens which drew over 2,000 guests.²⁸ This programme of social ostentation on the part of Dublin's élite class had no doubt less to do with the content of science *per se* than with the need of the Unionist community in Dublin to demonstrate its affiliation to enlightened British culture, and indeed beyond that to cultural cosmopolitanism, at a time of resurgent nationalism.

Two years later, the last IFCU Triennial Conference took place at Rosapenna in Donegal, attracting just forty-seven participants, mainly from the Belfast and Dublin clubs. By this time Praeger and Robert Welch considered that the IFCU had served its purpose, by introducing the members of the various clubs to each other's districts; the interdisciplinary focus provided by the joint excursions had led to the setting up of the working parties for the Clare Island Survey, at that time already under way. At the same time it is clear, however, that the passing of the IFCU resulted from the fact that the countrywide movement had quite simply spent the enthusiasm of its early years. The country had been visited, however cursorily, the more attractive plant and animal groups were tolerably thoroughly known, and it remained to the experts to continue with their work in the specialist fields. The demise of the IFCU, it should be emphasised, did not entail the demise of the field club movement as a whole: the Belfast tradition in particular would remain strong, and it is apparent that the club even acquired a new lease of life later, during the troubled times, when the BNFC became a focus for territorial loyalty in the beleaguered north-east. (In 1923, with BNFC membership at 571,²⁹ consideration was being given to limiting the membership to 600, since 'the rapidly increasing membership of the Club was becoming a rather embarrassing problem for the officials'.³⁰ The proposal was not adopted: BNFC membership peaked at 777 in 1924.)

Finally, there is one episode during the Donegal outing of 1910 which captures the spirit of the entire movement. On 12th July, the IFCU party set off on a trip to Tory Island, on board the SS *Cynthia*. This passenger steamer, which served Londonderry and Moville on Lough Foyle, had been procured at short notice, along with a special permit allowing the Tory Island trip, because the ship which had originally been chartered for the occasion had been damaged near Portrush and was being overhauled. The weather was favourable at the beginning of the twenty-four kilometre crossing to Tory past Horn Head, but by the time of arrival a dense fog had set in. Some breaks in the fog did allow for fieldwork and photography, but by the time the group embarked on the homeward trip, the fog was still obscuring everything. The steamer was now faced with the task

Science and the Cultural Revival: 1863–1916

of trying to crawl past the huge cliffs of Horn Head and to turn south-east into the calmer waters of Sheep Haven. The steamer eventually put down anchor at the base of Little Horn, south-east of Horn Head, when encroaching darkness made the prospect of continuing quite hazardous. Praeger's report continues as follows:

> Air and water alike were still, and the only sound was the incessant clamour of the birds – the musical cries of hundreds of Kittiwakes, the hoarse notes of Guillemots and Razorbills and the shrill piping of their young, and the calling of Herring Gulls. Time passed slowly, but presently, as darkness was falling, a cheer heralded the approach of a long white fishing-boat. From her crew the befogged party learned their position close in under the 'Little Horn', south-east of Horn Head; but in view of the gathering darkness and the heaviness of the fog, the captain decided not to move. So the party settled down for a night at sea. A few cushions and rugs were produced, and lifebelts were requisitioned as pillows. A smoking concert was organised on the upper deck, in which Mr. McDonald, assistant manager at Rosapenna, proved invaluable; and at 10.30 'dinner' was announced – a cup of tea without milk and one sandwich all round. By 1 a.m., all was silence, but a couple of hours later the birds again took up their chorus, and a new day came. At four o'clock our indomitable waiter went round with a number of lumps of sugar in a saucer – the last of the provisions. At seven, the fog seemed a trifle lighter, and the captain warily crept away eastward, and presently land was sighted which was made out to be Black Rock off Rosguill. Then the end came with startling suddenness. The mist began to lift; soon the sun came bursting through; and by 8 o'clock the 'Cynthia' came up to Downings Pier in full sunlight, with the mist rolling in sheets of flowing white off the surrounding hills. Never was breakfast more welcome than that to which the party sat down half an hour later.[31]

This incident exemplifies the innocence, bravura, and cheerful optimism of the field club members in the face of adversity. In the same report, Praeger remarked upon the proverbial good fortune of the IFCU with regard to weather on its triennial excursions; in retrospect, we can admire the debonair adventurousness of the field club members in their explorations of the Irish countryside. Their educated curiosity about Ireland's landscape led them to embrace the countryside in its fullest dimensions, antiquarian and scientific, and the decline of the latter to the advantage of the former was a real loss, from which we are only gradually beginning to recover.

The following decade was to bring about changes in the wider world to which natural history and the field clubs would not be immune. Wartime restrictions on movement meant that the exchange of people and information between Ireland and Britain as well as further exploration of the Irish countryside would be suspended for a long time. Victorian natural history participated in that optimistic and in some senses naive view of the world which respectable society held until the terrible caesura of the First World War. In Ireland, there was the further watershed of the 1916 Rising, which announced the rebirth of the phoenix of political violence and a phase of brutal contestation that left little room or relish for peacetime pursuits such as natural history. Many relevant dates register the sense of an era on the way out: 1915 saw the completion of the publication of the Clare Island Survey; in 1923 Praeger retired from his position as director of the National Library and would spend much of the following decade out of Ireland studying plant taxonomy; by that time many of the founding figures of the scientific movement of the 1880s and 90s were dead. Finally, in 1924, the *Irish Naturalist* ceased publication, having become a specialised journal of systematic biology without any firm subscription base.

NOTES

1. John Wilson Foster, 'Natural History, Science and Irish Culture', *Irish Review*, 9 (1990): pp. 61–9, and 'Natural Science and Irish Culture', *Eire-Ireland*, 26, no.2 (1991): pp. 92–103. Dorinda Outram, 'Negating the Natural: or Why Historians Deny Irish Science', *Irish Review*, 1 (1986): pp. 45–49.
2. Gordon L. Herries Davies, 'Irish Thought in Science', in Richard Kearney, ed., *The Irish Mind: Exploring Intellectual Traditions* (Dublin, 1985), pp. 294–310, p. 308.
3. *Ibid.*, p. 309.
4. Hilary Rose and Steven Rose, *Science and Society* (Harmondsworth, 1969), p. 25.
5. Steven Yearley, 'Colonial Science and Dependent Development: The Case of the Irish Experience', *Sociological Review*, 37, no. 2 (1989): pp. 308–331, pp. 319–20.
6. F.S.L. Lyons, *Ireland Since the Famine* (London, 1971), p. 201.
7. Terence Brown, *Ireland: A Social and Cultural History 1922–79* (London, 1981), p. 92.
8. See C.B. Moffat, *The Life and Letters of Alexander Goodman More*, with a preface by Frances M. More (Dublin, 1898).
9. Foster, 'Natural Science and Irish Culture', p. 95.
10. Robert Patterson, *The Natural History of the Insects Mentioned in Shakespeare's Plays* (London, 1841); *Introduction to Zoology, for the Use of Schools*, part 1 (London, 1846); *Introduction to Zoology, for the Use of Schools*, parts 1 and 2 (Belfast, 1849); *First Steps to Zoology* (Belfast, 1849).
11. See Terence Brown, *The Whole Protestant Community: The Making of a Historical Myth* (Derry: Field Day Theatre Co., 1985), pp. 22–23. The 1850s saw a revival of evangelical preaching in Ulster in open air settings quite distinct from normal church-going venues. Some of these aggressive anti-Catholic sermons were directly linked to sectarian street-rioting which followed. See David Hempton and Myrtle Hill, *Evangelical Protestantism in Ulster Society 1740–1890* (London, 1992), pp. 124 and 173.
12. R.A. Jarrell, 'The Department of Science and Art and Control of Irish Science, 1853–1905', *Irish Historical Studies*, 23 (1983): pp. 330–47. Lawrence W. McBride, *The Greening of Dublin Castle: the Transformation of Bureaucratic and Judicial Personnel in Ireland, 1892–1922* (Washington D.C., 1991), pp. 81–85.
13. See *Journal of the Co. Kildare Archaeological Society* (1891–), *Journal of the Cork Historical and Archaeological Society* (1892–), *Journal of the Waterford and South-East of Ireland Archaeological Society* (1894–), and *Journal of the Galway Archaeological and Historical Society* (1900–).
14. Praeger to Mary Isabella Leebody, 9th October 1892, National Botanic Gardens, Glasnevin.

15. G.W.D.Bailey, 'History of the Dublin Naturalists' Field Club', in *Reflections and Recollections: 100 Years of the Dublin Naturalists' Field Club* (Dublin, 1986), pp. 6–30.
16. *Irish Naturalist*, 3 (1894): p. 259.
17. *Ibid.*, p. 260.
18. 'The Irish Field Clubs 2: The Dublin Naturalists' Field Club', *Irish Naturalist*, 3 (1894): pp. 211–15, p. 211.
19. 'The Irish Field Clubs 3: The Cork and Limerick Naturalists' Field Clubs', *Irish Naturalist*, 3 (1894): pp. 247–52, p. 247.
20. 'The Irish Field Clubs 1: The Belfast Naturalists' Field Club', *Irish Naturalist*, 3 (1894): pp. 141–45, p. 141.
21. 'The Cork and Limerick Naturalists' Field Clubs', pp. 250–51.
22. *Ibid.*, p.252.
23. *Ibid.*, p.252.
24. H.A. Miers, 'The Educational Opportunities of Local Scientific Societies', *Irish Naturalist*, 17 (1908): pp. 215–18, p. 216. *Report of the Seventy-Eighth Meeting of the British Association for the Advancement of Science, 1908* (London, 1909), pp. 917–27. See also Rose and Rose, *Science and Society*, Chapter 2.
25. *Handbook to the City of Dublin and the Surrounding District, Prepared for the Meeting of the British Association*, ed. G.A.J. Cole and R. Ll. Praeger (Dublin, 1908).
26. *Daily Express*, 4th September 1908, p. 12.
27. *Daily Express*, 4th September 1908, p. 12 and 5th September 1908, p. 8.
28. *Irish Naturalist*, 17 (1908): pp. 247–48.
29. *Proceedings of the Belfast Naturalists' Field Club*, (2)8 (1922–3): p. 213.
30. *Proceedings of the Belfast Naturalists' Field Club*, (2)8 (1923–4): p. 301.
31. 'General Account of the Sixth Triennial Conference and Excursion Held at Rosapenna, July 8–10, 1910', *Irish Naturalist*, 19 (1910): pp. 157–66, pp. 163–64.

Index of People and Institutions

Academie des Sciences, Paris, 87
Aher, David, 68
Airy, George B., 19, 23, 27
Alexander, Prof., 95
Allen, David, 131
Anderson, Benedict, 10
Andrews, John, 119, 129
Andrews, Thomas, 90
Apjohn, Prof., 93
Armagh History and Philosophical Society, 42
Armstrong, Edward A., 121, 131
Arnold, Thomas, 130–1
Attwood, Benjamin, 78

Babbage, Charles, 19–20
Bacon, Francis, 22
Bald, Robert, 80
Ball, John, 38
Ball, R.S., 39
Banks, Sir Joseph, 75, 77
Banville, John, 130
Basalla, George, 4, 11, 50
Bateman, James, 74
Beckett, J.C., 51
Belfast Municipal College of Technology, 95–6
Belfast Natural History and Philosophical Society, 156
Belfast Naturalists' Field Club, 156, 159, 162
Bentham, Jeremy, 25, 27
Bhabha, H., 8
Birmingham, John, 39
Blache, Vidal de la, 137
Blacker, William, 104
Blacke, Sir Valentine, 160
Boole, George, 44
Botanic Gardens, Glasnevin, 51, 56, 157
Brewster, David, 21–3
British Association for the Advancement of Science, 19, 21, 22–4, 26, 30–1, 39, 51, 161

British Mineralogical Society, 73
Brougham, Henry, 21, 23
Brown, Terrence, 49
Buddle, J., 77
Burgoyne, J.F., 90
Burns, W.L., 12
Butler, Eleanor, 145

Callan, Nicholas, 38, 90, 123
Calman, W.T., 57
Cambrensis, Giraldus, 121
Cambridge University, 24, 27–8
Campbell, John, 142
Carpenter, George Herbert, 53–4, 56, 61, 158
Casey, John, 38
Catholic University. *See* University College Dublin
Chakrabarty, D., 8, 11
Civil Engineers' Society of Ireland, 90
Clampett, A.G.T., 142
Clare Island Survey, 155
Clarke, Desmond, 119, 122, 126
Cochrane, A. (Lord Dundonald), 71, 72
Cohen, I. Bernard, 4
Cole, Grenville Arthur James, 56, 58–9, 60, 62–3, 137–9, 161
Cole, Henry, 108
Coleridge, Samuel Taylor, 20, 24–5, 26, 128
Colgan, Nathaniel, 155
Cooper, E.J., 39
Corcoran, Timothy Joseph, 52, 142–4
Corry, Thomas, 131
Cosslett, Tess, 125, 126
Cullen, L.M., 86
Curwen, J.C., 73–4
Curzon, Lord, 12

Davies, G. Heries, 119, 136, 154, 156
Davis, Thomas, 122–3, 147
Davis, William Morris, 138

167

Davy, Sir Humphry, 77, 128
De Morgan, Augustus, 11, 23
Department of Agriculture and Technical Instruction, 54–5, 56, 57, 58, 109, 140, 141–2, 143
Department of Science and Arts (London), 39, 89, 109–110, 151
De Vere, Aubrey, 24, 27, 29
Doyle, Martin, 104
Drummond, Thomas, 38, 39
Drummond, William Hamilton, 130
Dublin Museum of Science and Art. *See* National Museum
Dublin Naturalists' Field Club, 158–9
Dublin Society. *See* Royal Dublin Society
Dublin Technical School, 112
Dupin, C., 88
Durham University, 86

Eager, A.R., 119
Ecole Centrale des Arts et Métiers, 87, 88
Ecole Polytechnique, 87, 88, 91
Ecole des Ponts et Chaussees, 87
Edgeworth, Richard Lovell, 69–70, 76
Erck, Wentworth, 39
Evans, Estyn, 146

Faraday, Michael, 78–9
Farey, John, 71, 73, 74–5
Farming Society of Ireland, 104
Farran, G.P., 57
Fereday, Samuel, 77
Fitzgerald, George Francis, 51, 95
Fleming, Donald, 4
Fletcher, George, 142
Forster, E.M., 12
Foster, John Wilson, 49, 153, 155
Foucault, Michel, 10
Freeman, W.T., 145, 146
Friel, Brian, 129
Fresnel, Augustin, 19, 20

Geerz, Clifford, 148
Geological Survey of Ireland, 51, 56, 90, 107, 136–7
Gill, T.P., 58, 61, 109
Glasnevin Model Farm, 105
Goldsmith, Oliver, 130
Gosse, Edmund, 126
Grand Canal Company of Ireland, 68, 72
Gray, William, 131

Green, William Spotswood, 121
Greenough, G.B., 75
Griffith, Richard, 38, 71, 136
Grubb, Thomas, 96
Grundtvig, Bishop, 140–1

Haddon, Alfred Cort, 158
Halbert, James Nathaniel, 158
Hales, William, 39
Hallissy, Timothy, 139
Hamilton, Sir William Rowan, 2, 19–31
Hankins, Thomas, 20
Harcourt, Sir William, 13
Hargreave, C.J., 38–9
Hart, Henry Chichester, 131
Haughton, Samuel, 42–3
Heaney, Seamus, 130
Hennessy, Henry, 38, 42, 43, 44, 45
Hennessy, J.P., 38, 43
Henry, R.M., 52
Herschel, J.F.W., 20, 23
Heron-Allen, E., 60–1
Higgins, William, 38
Hodges, John, 105, 115
Hodgson, Rev. J., 77
Holt, E.W.L., 55, 56, 57–8
Hughes, T. Jones, 146, 147
Huxley, T.H., 131
Hutchinson, Clive, 121
Hyde, Douglas, 127, 155

Irish Agricultural Organisation Society, 106

Jarrell, Richard A., 51, 123, 124
Johnson, Henry, 79
Joly, John, 39, 51
Joyce, James, 122, 129, 155

Kane, Sir Robert, 38, 41–2, 43, 44–5, 51, 90, 91, 107–8, 115, 120
Kearney, Richard, 154
Keenan, Sir Patrick, 111
Kelly, John, 104
Kirwan, Richard, 39, 68
Knowles, Matilda, 158
Kropotkin, P., 135, 138
Kumar, Deepak, 6

Larmor, Joseph, 39
Latour, Bruno, 5, 10
Lee, Joe, 49
Lloyd, Bartholomew, 90, 91

Index 169

Lloyd, Humphrey, 19, 20, 23, 91–2, 96–7
London University, 88
Luby, Thomas, 91
Lyons, F.S.L., 49–50, 155

McCoy, Frederick, 62
McCullagh, James, 39, 91
MacDonagh, Oliver, 2
MacDonnell, Sir A., 61
MacDonnell, James, 72
McGee, D'Arcy, 130
Mackinder, Halford, 137
MacLeod, Roy, 50–1
MacNeill, Eoin, 52–3, 142–3
MacNevin, W.J., 39
McWeeney, Dr., 151
Maine, Sir Henry, 12
Mallett, Robert, 90
Maskelyne, Neville, 29
Matthews, William, 79
Maynooth. *See* Saint Patrick's College, Maynooth
Mechanics' Institutes, 26, 89, 107
Miers, Prof. H.A., 161
Mill, John, 25
Mill, John Stuart, 25, 27
Moloney, Michael, 127–8
Molloy, Gerald, 38
Molyneux, William, 130
Monteagle, Lord, 112
Moore, David, 155
More, Alexander Goodman, 155
Morrell, Jack, 24, 39
Morris, Meaghan, 11
Mostyn, Sir Thomas, 68
Munster Institute, 115
Murphy, Edmund, 104, 105
Murphy, J.J., 131
Murphy, J.N., 104–5
Murray, John, 78
Museum of Irish Industry, 39, 42, 45, 58, 93, 107–8, 115, 137

National Academy, 52
National Library, 54, 157
National Museum of Ireland, 51, 53, 56, 57–8, 59–61, 62–3, 155–6
National University of Ireland, 52, 143
Natural History Museum (London), 57, 59—61
Newman, John Henry, Cardinal, 52, 108
Nichols, Albert Russell, 158
Nimmo, Alexander, 39

O'Connell, C.S., 145, 146–7
O'Connell, Daniel, 30
O'Connor, Bernard, 123
O'Connor, Frank, 128
O'Curry, Eugene, 130
Ogilvie, F.G., 59
O'Grady, S.J., 127
O'Hara, J.G., 20
Oldham, C.H., 142
Ophir, Adi, 9, 121, 124, 128
Osborn, Michael, 11, 13
Outram, Dorinda, 153

Palmer, H.R., 89
Parsons, Charles, 97
Patterson, Robert, 131, 156
Patterson, Robert Lloyd, 156
Pearse, Padraic, 147–8
Perry, Samuel, 70, 72
Phillips, William, 73
Playfair, Lyon, 108
Plücker, Julius, 20
Plunkett, Sir Horace, 52, 95, 106, 109, 139–42, 144
Portlock, J.E., 58
Potter, Richard, 21
Praeger, Robert Lloyd, 54, 56, 59–60, 121, 154–64
Preston, Sir J., 111
Preston, T., 112

Queen's Colleges, 39, 41, 42–3, 46, 91, 92, 105, 108, 111, 137
Queen's College Cork, 41, 42, 44, 111, 115, 145, 146, 147
Queen's University of Belfast, 95–6, 105, 145, 146

Ramsden, Jesse, 29
Rathmines School of Commerce, 142
Rhodes, Israel, 68
Richards, I. Van W., 70
Robinson, Thomas Romney, 39, 42
Robison, John, 22
Roosevelt, Theodore, 140
Rosse, Earl of, 39, 44, 51, 96
Royal College of Science of Ireland, 39, 51, 53, 93, 96, 138
Royal Dublin Society, 39, 51, 69, 90, 103–4, 107, 136
Royal Institution, 78
Royal Irish Academy, 29, 40, 51, 52, 123, 161
Royal Military Academy, Woolwich, 88, 93

Russell, George, 148
Ryan, James, 67–80

Saint Patrick's College, Maynooth, 52, 126
Salmon, George, 39
Sarton, George, 20, 30
Scharff, Robert Francis, 56, 57–8, 61, 62–3, 158
Shapin, Steven, 9, 121, 124, 128
Shaw, George Bernard, 129, 148
Smith, William, 75
Society of Arts, 77
Society for the Diffusion of Useful Knowledge, 26
Southern, Rowland, 158
Sowerby, James, 75–6
Sprat, Thomas, 41
Stephenson, George, 77
Stewart, S.A., 131, 155
Stokes, George Gabriel, 20
Stoney, George Johnstone, 39
Strachan, A., 59
Sullivan, William K., 38, 40, 42–3, 45–6, 108, 111, 115
Swanston, William, 60–1
Synge, J.M., 129–30, 155

Technical Schools, 95
Telfer, William, 71
Thackray, Arnold, 24, 39

Thompson, William, 121, 131
Thomson, James, 39
Trinity College Dublin, 19, 21, 28–9, 41, 42–3, 51–2, 90, 91, 93, 95–6, 108, 146
Tyndall, John, 51, 125, 126

University College Dublin, 43–5, 51, 52, 92–3, 108, 139, 142, 147
Vallencey, C., 69
Vaughan, W.E., 105
Vignoles, C.B., 88
Viney, Michael, 131

Wallerstein, Immanuel, 5
Ward, Margaret, 50
Ward, W. (Viscount Dudley and Ward), 76–7
Welch, Robert, 159
West, C.D., 96
Whewell, William, 19, 22–3, 25, 26
Wilde, Oscar, 129
Williams, Raymond, 128
Wordsworth, William, 24, 26
Wright, Joseph, 59–61
Wyndham, George, 61
Wyse, Thomas, 90–1

Yeats, William Butler, 124, 127, 129, 130